Biodiversity and Conservation

The value of planet Earth's biodiversity has been estimated at US$16–54 trillion per year. Trillions of dollars of food, raw materials, pharmaceuticals, oxygen production, soil nutrient enrichment, climate regulation and sheer aesthetic delight and wonder. We are reliant on biodiversity to keep the planet healthy and resilient.

The second edition of *Biodiversity and Conservation* continues to offer an introductory guide through the maze of interdisciplinary themes that combine under the concept of 'biodiversity'. Using engaging examples throughout, the text combines biological sciences with its insights into the origins, variety and distribution of biodiversity, with the analysis of the social and political context, the threats to and opportunities for the survival of natural systems. Whilst retaining its existing structure the new edition reflects advances that have demonstrated the importance of living systems as drivers of environmental services vital to human health and security. The processes driving the creation and distribution of biodiversity have been updated to reflect new research. The final chapter has been revised to tackle more explicitly the contrasting approaches to conservation.

The text remains the only introductory book bringing together the full range of science and social sciences, theory and practice that goes to make up biodiversity and conservation.

Michael J. Jeffries is Senior Lecturer in Ecology at Northumbria University.

Routledge Introductions to Environment Series

Published and Forthcoming Titles

Routledge Introductions to Environment Series

Biodiversity and Conservation

Second edition

Michael J. Jeffries

Routledge
Taylor & Francis Group

LONDON AND NEW YORK

First published 1997
by Routledge

The second edition published by Routledge 2006

2 Park Square, Milton Park,
Abingdon, Oxon OX14 4RN

Simultaneously published in the USA and Canada
by Routledge
270 Madison Ave, New York, NY 10016

Transferred to Digital Printing 2009

Routledge is an imprint of the Taylor and Francis Group, an informa business

Typeset in Times New Roman and Franklin Gothic
by Keystroke, Jacaranda Lodge, Wolverhampton
Printed and bound in Great Britain
by TJI Digital, Padstow, Cornwall

British Library Cataloguing in Publication Data
A catalogue record for this book is available from the British Library

Library of Congress Cataloging in Publication Data
Jeffries, Mike J., 1958–
Biodiversity and conservation / Michael J. Jeffries.— 2nd ed.
    p. cm. – (Routledge introductions to environment series)
    Includes bibliographical references and index.
    ISBN 0–415–34299–6 (hardcover) – ISBN 0–415–34300–3 (softcover)
    1. Biological diversity. 2. Biological diversity conservation.
    I. Title. II. Series.
    QH541.15.B56J44 2005
    333.95–dc22   2005001109

ISBN10: 0–415–34299–6 (hbk)
ISBN10: 0–415–34300–3 (pbk)

ISBN13: 9–78–0–415–34299–5 (hbk)
ISBN13: 9–78–0–415–34300–8 (pbk)

Contents

Series editors' preface
Environmental Science titles

The last few years have witnessed tremendous changes in the syllabi of environmentally related courses at Advanced Level and in tertiary education. Moreover, there have been major alterations in the way degree and diploma courses are organised in colleges and universities. Syllabus changes reflect the increasing interest in environmental issues, their significance in a political context and their increasing relevance in everyday life. Consequently, the 'environment' has become a focus not only in courses traditionally concerned with geography, environmental science and ecology but also in agriculture, economics, politics, law, sociology, chemistry, physics, biology and philosophy. Simultaneously, changes in course organisation have occurred in order to facilitate both generalisation and specialisation; increasing flexibility within and between institutions is encouraging diversification and especially the facilitation of teaching via modularisation. The latter involves the compartmentalisation of information which is presented in short, concentrated courses that, on the one hand are self-contained but which, on the other hand, are related to prerequisite parallel and/or advanced modules.

These innovations in curricula and their organisation have caused teachers, academics and publishers to reappraise the style and content of published works. While many traditionally styled texts dealing with a well-defined discipline, e.g. physical geography or ecology, remain apposite there is a mounting demand for short, concise and specifically focused texts suitable for modular degree/diploma courses. In order to accommodate these needs Routledge has devised the Environment Series which comprises Environmental Science and Environmental Studies. The former broadly encompasses subject matter which pertains to the nature and operation of the environment and the latter concerns the human dimension as a dominant force within, and a recipient of, environmental processes and change. Although this distinction is made, it is purely arbitrary and for practical rather than theoretical purposes; it does not deny the holistic nature of the environment and its all-pervading significance. Indeed, every effort has been made by authors to refer to such interrelationships and provide information to expedite further study.

This series is intended to fire the enthusiasm of students and their teachers/lecturers. Each text is well illustrated and numerous case studies are provided to underpin general theory. Further reading is also furnished to assist those who wish to reinforce and extend their studies. The authors, editors and publishers have made every effort to provide a series of exciting and innovative texts that will not only offer invaluable learning resources and supply a teaching manual but also act as a source of inspiration.

A. M. Mannion and Rita Gardner

Series International Advisory Board

Australasia: Dr Curson and Dr Mitchell, Macquarie University

North America: Professor L. Lewis, Clark University; Professor L. Rubinoff, Trent University

Europe: Professor P. Glasbergen, University of Utrecht; Professor van Dam-Mieras, Open University, The Netherlands

Note on the text

Bold is used in the text to denote words defined in the Glossary. It is also used to denote key terms.

Plates

Figures

Tables

Boxes

Author's preface

April 2004 – and I was marooned in Scotswood Natural Community Garden in the West End of Newcastle upon Tyne, north-east England, the minibus due to give me a lift home having vanished. Newcastle is a top ten global party city according to Condé Nast's *Traveller* magazine with a mile of glittering pubs and clubs along its Quayside but Scotswood, barely ten minutes stroll away, may as well be on another continent as far as the party city revellers are concerned. Yet packed into this tiny garden in the heart of one of the UK's most deprived areas you can find a wealth of crops grown by permaculture used to promote healthy eating, a reed bed system used to filter and clean the streams, burrowing bees dancing across sunny banks of wild flowers, a dragon woven from willow stems through which children can crawl and one of the best ponds for miles around, teeming with plants and minibeasts. The garden has hosted workshops on traditional crafts, celebrated the apple and used an outdoor oven to bring in disparate and sometimes mutually suspicious communities to build bridges. Earth care, people care and fair share as their website puts it. As for me – I wasn't quite as stuck as it seemed because the No. 38 bus into the city centre stops nearby and there was the Garden's homemade apple and walnut cake to keep me going. Food, wildlife, social equity, environmental quality, health, conservation. You do not have to go to the heart of the Amazon rainforest or the Serengeti to find the significance of biodiversity in all our lives.

I hope this second edition continues the mystery, excitement and possibilities that are bound up in the science of biodiversity. As with the first edition this will be little thanks to my expertise in the field (I could tell you about pond life and the strange history of wildlife on TV but that's about it) and again this edition is a pillaging of other people's insights, knowledge and energy. I hope those whose work I have ransacked find some consolation that their names, ideas and publications cited in this book will inspire others in the same way as they have me. Students should use this book as a springboard to dive into the wider literature.

I am very grateful to those who have helped with the production of this book. For permission to use pictures Daphne Christelis (Greenpeace), Dave Clarke (Zoological Society of London), Glenys Dawkins (NERC Centre for Population Biology), Chris Gibbins (Aberdeen University), Annabelle Lea (Durrell Wildlife Conservation Trust), Rob Morley (somewhere in Africa) and Martin Woodcock. To Andrew Mould, Adam Gilbert and especially Zoe Kruze at Routledge for all their patient support; how Zoe put up with the only author not to own a PC or laptop I probably do not want to know.

Michael Jeffries
Newcastle

The Rio Earth Summit

June 1992. Representatives from over 180 governments met in Rio de Janeiro for the Earth Summit, the second United Nations Conference on Environment and Development.

Twenty years earlier the first UN conference on the Human Environment met in Stockholm. Widely pilloried as a dialogue of the deaf, the Stockholm conference fractured as developed countries asked poorer nations to clean up environmentally destructive development whilst the developing countries wished for economic growth, even if pollution and degradation were the price. The 1992 Rio Earth Summit risked a similar schism, with the developed world focusing on climate change, destruction of tropical forests and species loss, the developing world still desperate for economic improvement, but since Stockholm, two themes had emerged encouraging global empathy. First, in 1987 the Brundtland Report provided a defining moment, partly in response to the Stockholm fall out, promoting the concept of **sustainable development**. Second, the word **biodiversity**, scarcely heard of a decade before Rio, had a gained a global audience.

The Earth Summit spawned four major agreements: the Rio Declaration on Environment and Development (citing the rights and responsibilities of individual states), the Convention on Climate Change, Agenda 21 (wide-ranging objectives and approaches for sustainable development), and the Convention on Biological Diversity. This convention was signed by 155 states at the Summit between 5 and 14 June 1992. Of the 105 princes, presidents and prime ministers who had come to Rio (despite much touted threats from gangs of muggers and transvestite carnivals) and gave formal speeches, 67 specifically mentioned biodiversity (or biological diversity). Biodiversity had become a dominating theme, most famously immortalised by US President George Bush Senior. Faced with impending presidential elections Bush would not commit himself to attend, unwilling, as he put it, to save squirrels if it cost one American job. He did attend Rio, did not sign the convention whilst there, and did not get re-elected. However, his successful rivals had maintained a strong environmental agenda. The environment was a political issue.

A global convention, havering US president and 'biological diversity' a popular buzz-word for 67 heads of state – biodiversity had gained a familiarity greater than any other ecological concept and in a remarkably short space of time. This spectacular debut suggests something more than scientific rigour and academic interest. Biodiversity had caught a wider mood.

The junket in Jo'burg

September 2002. Representatives from over 180 governments, 100 heads of state, 22,000 accredited delegates and an estimated 40,000 other participants met at the UN World Summit on Sustainable Development, essentially a ten-years later follow-up to the Rio Earth Summit. 'People, Planet, Prosperity' was how the conference website tagline summarised the intent.

Ten years earlier the Rio Summit topped a crescendo of environmental concern that had been growing since the early 1970s. However, by Johannesburg the optimism of Rio had long gone. The official Johannesburg website acknowledged that 'progress in implementing sustainable development has been extremely disappointing since the 1992 Earth Summit'. The UK press coverage picked up on this theme of dashed hopes or focused on the apparent mismatch between hospitality for the official delegates versus the living conditions of South African shanty dwellers a kilometre down the road. The UK tabloid newspaper the *Sun*, under the banner headline 'It makes you sick', explored the catering with a list that resembled a biodiversity audit: 5000 oysters, 100 lb of lobsters and shellfish, 4000 lb of steak and chicken, 450 lb of salmon and 220 lb of kingclip (a South African fish delicacy). The event proved fraught with bitter demonstrations outside and the US Secretary of State, Colin Powell, being booed by delegates for suggesting that US economic polices benefited biodiversity. Greenpeace produced a school report style assessment of the event comparing what was required with what was achieved: genetic engineering? Caught cheating. Toxics? Sickening performance. Oceans? Too much fishing for compliments and not enough real work. Friends of the Earth were more damning, concluding that the events finished up with a stunning lack of progress.

You would expect political heat and noise but the degree of cynicism and sense of failure in both the run-up and aftermath of the event suggest a serious failure since the heady days of Rio. Many speeches, reports and agreements focused on the need to implement treaties and policies, not just rack up more and more international accords based on hollow good intentions.

In the midst of the Bollinger and booing, what had become of biodiversity as a major theme? Biodiversity had not been forgotten. The Johannesburg summit was organised around five key themes, the so-called WEHAB Initiative championed by the UN as priorities, especially for the world's poor – Water, Energy, Health, Agriculture and Biodiversity. Biodiversity was demonstrably not solely a branch of the biological sciences but recognised as the foundation for sustainable development. Conference sessions inextricably linked biodiversity with economics and development, for example the **World Conservation Union's (IUCN's)** 'Poverty and Biodiversity Day' and 'Business and Biodiversity Day' on 28 and 30 August respectively.

If it is no longer easy to forge international treaties on biodiversity or carry out the resulting obligations this is because the variety and richness of life on earth is no longer the preserve solely of scientists tucked safely out of the way in laboratories or of natural history TV audiences enchanted by strange creatures in faraway lands. Biodiversity has become centre stage in the most complex and fraught social, political and economic challenges that we face. By 2002 biodiversity had an explicit financial value – as the World Summit on Sustainable Development put it, US$33,000,000,000,000.

① Biodiversity: from concept to crisis

Biodiversity is a recent concept, life on Earth a very ancient phenomenon. The history of both have fuelled awareness of a global environmental crisis. This chapter covers:

- **Origins of the term biodiversity**
- **Themes embraced by concept of biodiversity**
- **History of life on Earth**
- **Current biodiversity crisis**

The word **biodiversity** was coined in the mid-1980s to capture the essence of research into the variety and richness of life on Earth. The word is now widely used, its rapid establishment in science and popular culture an indication of the importance of the topic but also a source of confusion. First, biodiversity is the richness and variety of life on Earth. The flowers and insects and bacteria and forests and coral reefs are biodiversity. Second, biodiversity is an area of scientific research, including both description and measures of diversity and explanations of how this diversity is created. Biodiversity has been increasingly used as a conceptual focus for conservation policy and practice in response to one of the strongest themes underpinning the founding work on biological diversity, species extinction and ecosystem loss, brought to global prominence by the Rio Earth Summit. The variety of life on Earth can be investigated at different levels: genetic variation, the numbers of species, the extent of ecosystems. Nature's creative and destructive forces include ecological and evolutionary processes. Current degradation of biodiversity is driven by human pressures and conservation responds with protective laws, reserves and refuges in captivity. The purpose of this book is to provide an introduction to all these topics. Chapter 1 outlines the history of biodiversity, both as a concept and the story of life on Earth up to the current crisis. Chapter 2 describes the ecological and evolutionary processes that spawn this diversity and govern its distribution. Chapter 3 provides an inventory of how biodiversity can be classified and quantified. Chapter 4 describes recent extinctions and their causes, natural and human. Chapter 5 outlines policy and practice to conserve biodiversity.

First then, how did a word which was unheard of in 1980 come to be the heart of a United Nations global treaty only twelve years later?

The road to Rio: the conceptual history of biodiversity

Origins

The diversity of life on Earth has been a central theme of the natural sciences but touches on many other areas. The Bible credits Adam with the job of naming the animals, a fundamental task for facing those quantifying biodiversity. The same approaches to classifying life are apparent in ancient and modern societies. Western culture has repeatedly revised its understanding of the variety and nature of life. The Greek philosopher Aristotle recognised between 500 and 600 species, echoing modern folk classifications which typically recognise 300 to 600. Slavish copying of classical texts was abandoned during the sixteenth and seventeenth centuries, spurred by technological advances and the spread of ideas through printing. Classification increasingly focused on the species. The nineteenth century saw the final abandonment of the folk biology principle that lumped species together by broad type (e.g. tree) in favour of biodiversity described by detailed structure and relatedness. Our attitudes to life continue to change: the twenty-first century is dominated by our understanding of evolution, which is so powerful a theory that, according to Richard Dawkins, it has rendered God superfluous. At the same time ancient patterns linger; the common use of 'r' and 'l' sounds in words for 'frog' suggesting an intuitive, onomatopoeic appreciation of how we describe living things. The richness of life has been central to human society and science but the term biodiversity is an upstart.

The origins of term are credited to two papers published in 1980 (Lovejoy 1980; Norse and McManus 1980). Lovejoy, working for the World Wildlife Fund in Washington DC, contributed to *The Global 2000 Report to the President* of the United States, reviewing global environmental topics such as energy, human population and economics. Examining the extent of global forestry resources, Lovejoy reviewed two consequences of forest exploitation: changes to global climate and to biological diversity. Estimates of extinctions based on different forest loss rates were given. Lovejoy wrote of biological or biotic diversity defined as simply the total number of species. He wrote of biological capital, the inability of markets to properly value ecological systems and the functions that ecosystems fulfil that are of benefit to people.

Norse and McManus were ecologists on the White House Council on Environmental Quality during Jimmy Carter's presidency and in 1980 contributed a chapter to the *Eleventh Annual Report of the Council on Environmental Quality* entitled *Ecology and Living Resources: Biological Diversity*. The chapter examines global biodiversity which is defined as two related concepts, genetic and ecological diversity (the latter equated with numbers of species). The bulk of the chapter discusses the material benefits of biological diversity, psychological and philosophical bases for preservation, human impacts that were losses and strategies and policy for conservation.

The context of these papers is important. Biodiversity was discussed on a global scale. The bulk of the work dealt with wider themes than the purely biological. The importance of biodiversity, actual and potential, was apparent. There is a recognition that the activity of natural ecosystems provides what are now called services or functions vital to a healthy planet. There is a political and philosophical slant to some of the discussion. This usage carries much of the resonance familiar nowadays, combining the richness of life, an awareness of loss, the importance of biodiversity for economics and an ethical, social dimension. Biodiversity is not solely a branch of biology.

The snappy abbreviation biodiversity (fleetingly BioDiversity) is credited to Walter Rosen, working for the American Natural Research Council/National Academy of Sciences, as a co-director for the 1986 Conference 'The National Forum on BioDiversity', held in Washington, DC. The publication of papers from the Forum, in a book entitled *BioDiversity* (Wilson 1988), was the spark igniting wider interest and usage. The contents of *BioDiversity* are revealing. The themes of the fifty-seven chapters extend beyond pure science. Six main threads reoccur. Eleven chapters provide inventories and estimates of the extent and types of life on Earth. Ten explore loss rates, historic and recent. Ten examine the **value**, financial, aesthetic and for planetary health, both as current uses plus future potential. Eleven analyse **economics**, financial structures, forces and approaches

Plate 1 The O'Hara farm, Galway. The O'Hara family abandoned their farm during the Irish Potato famine. Mr O'Hara ended up a match-seller on the streets of New York and never saw his children again. The failure of potato crops was due to dependence on one genetic variety vulnerable to disease. The lack of genetic biodiversity cost the O'Hara family dear.

especially as agents of destruction. Twenty-eight chapters discuss the policy, practice or examples of **conservation**. Eleven chapters range much more eclectically through social systems, religion, poetry and environmental politics, reflecting **attitudes** to biodiversity. E.O. Wilson, the editor, ends his introductory chapter with the suspicion that the success of conservation will be decided by an ethical response, that 'the fauna and flora of a country will be thought part of the natural heritage as important as its art, its language and that blend of achievement and farce that has always defined our species' (see Plate 1).

In 1997 *Biodiversity II* was published, a ten years down the line report on progress (Reaka-Kudla *et al.* 1997). There are revealing changes in the content. The thirty-three chapters are dominated by inventories, estimates and classifications of richness but often with strong themes of ecological and evolutionary processes running through time and space. Many of the studies are very detailed, for example of bower birds and snout moths. Microbial and genetic biodiversity are much more to the fore. There is much less on valuation, loss rates and economics, at least as separate themes, but more details of the ecosystems, species and processes. There are no poems either. The development of institutional infrastructure is prominent and seen as one of the more important messages, reflecting a maturing scientific discipline.

The wide scientific, social and philosophical horizons incorporated by the concept of biodiversity are evident in much of the early literature (see Table 1.1). The linkage of economic and political issues with biological science was important. Conservationists could point to the real financial value of biodiversity and conservation widened its horizons to address the underlying human pressures threatening life on Earth.

The IUCN had promoted the idea of a global convention on biodiversity since 1981 and in 1987 the **United Nations Environment Programme (UNEP)** called for

Table 1.1 Major themes within biodiversity: the presence of six main themes in texts from the conceptual origin of biodiversity in 1980

Theme	Lovejoy 1980	Norse and McManus 1980	Norse et al. 1986	Wilson 1988	Flint 1991	Reakka-Kudla et al. 1997
Inventory	Definition; global species total	Definition; global species total	Definition; global species total	Definition; global species total; habitat-specific totals	Definition; global species total	Definitions, genetic, species and ecosystem level; linking ecosystem processes to variety and richness
Loss rates	Estimated losses due to forest destruction	Causes (settlement, transport, fragmentation, agriculture, forestry, over-exploitation, introduced species)		Global and habitat-specific estimates; geological patterns case studies	Habitat examples; causes (development, market failure, interventions, habitat loss, over-exploitation)	Estimated global extinction rates, species and site-specific case studies
Value	Potential uses	Potential uses (food, energy, chemicals, raw materials, medicine)	Products (actual and potential); ecosystem services; forestry; psychological well-being	Medicine, industrial, food, potential	Use and non-use values, potential, habitat examples	Specific examples of value to biotechnology and agriculture; value as a scientific tool and to raise awareness
Economics	Valuation of ecosystem function		Timber production	Market failure to price biodiversity, new economic approaches, pricing biodiversity	Economic approaches to value biodiversity	

Table 1.1 continued

Theme	Lovejoy 1980	Norse and McManus 1980	Norse et al. 1986	Wilson 1988	Flint 1991	Reakka-Kudla et al. 1997
Conservation		Gene banks, botanical gardens, species specific schemes, ecosystem management, legislation	Concepts; conservation and management of forests and rare species; laws	Priorities, case studies, technologies, reintroductions	Aid policy procedures for priority sites, habitat examples, criteria	Strategic and case-specific examples, marine and terrestrial; genetic, species and ecosystem level; emphasis on infrastructure
Attitudes		Philosophical, psychological, affinity to life, right to exist	Ethics and stewardship	Green movement, society links to ecosystem, morals, religion, philosophy	Ethical value of biodiversity	Brief review of history in science

an international treaty to govern conservation and sustainable use of biodiversity. The years of preliminary meetings reflected political tensions. Developing countries, many harbouring the greatest concentrations of biodiversity, were keen to retain full control and rights to exploit these resources. Some developed countries promoted an awkward mixture of biodiversity as everyone's global heritage, over which the impecunious developing world was expected to stand guard for all our sakes, coupled with a desire that industries should have free rein to explore and develop new products from the treasure trove. Developing countries were fearful that any discoveries would be whisked away to foreign labs and reappear, maybe in synthetic form, ringed about by strict patent controls.

The Rio Convention consists of forty-two articles, covering definitions, principles, research, education, finance and administration. Article 1 summarises its objectives:

> The objectives of this Convention, to be pursued in accordance with its relevant provisions, are the conservation of biological diversity, the sustainable use of its components and the fair and equitable sharing of the benefits arising out of the utilisation of genetic resources, including the appropriate access to genetic resources and by appropriate transfer of relevant technologies, taking into account all rights over those resources and to technologies, and by appropriate funding.

Post Rio

The Convention on Biological Diversity (commonly abbreviated to CBD) is now institutionalised as a UN-based organisation with its own extensive secretariat and primarily answering to the signatory countries who meet as the Conference of Parties (COP). The CBD has its own website providing thorough access to its structures, activities and related sites: http://www.biodiv.org. As of mid-2005 there were 188 parties to the Convention, 168 of them having ratified the treaty and therefore obliged to implement the agreement nationally. The CBD website reveals the wide domain of biodiversity as a theme, the site header containing the tagline 'conservation, sustainable use, equitable benefit sharing'. The CBD Secretariat is keen to promote this wide remit, for example the logo has been described as illustrating biodiversity's significance underpinning security of food, water and health.

Driving the implementation of the Rio Convention has not been easy. Progress has been directed by meetings of the COP. The second Conference of Parties proposed a Global Biodiversity Outlook as a tool to allow countries to review progress, identify problems, set priorities and communicate. The first version was published in 2001, but at the Johannesburg summit Peter Schei, a lead adviser to the UN, reported that 'there had not been a lot of implementation' and described as ridiculous the amount spent on conserving biodiversity compared to all the subsidies that typically drive the loss and degradation of natural systems. The Johannesburg summit was dominated by the need for action rather than words. Summarising actions related to biodiversity the UN Secretariat emphasised the need to

- develop mechanisms for concrete action
- incorporate biodiversity as a mainstream aspect of global and local economics and trade
- share results, technology, good practice and build the capacity for conservation, especially at the local level in the developing world
- address poverty as a major threat to biodiversity
- ensure equitable benefits from biodiversity, especially through property rights.

Johannesburg resulted in the Plan of Implementation, a catch-all agreement requiring governments to set out detailed action plans to meet their commitments to existing UN conventions, such as the Rio Convention on Biodiversity.

The Sixth CBD Conference of Parties met in 2003 and emphasised the need to embed the conservation of biodiversity into wider sustainable development. The UN and leading NGOs convened 'Biodiversity after Johannesburg' to highlight the role of biodiversity and ecosystem services in meeting Millennium Development Goals (MDGs). These goals were set by the UN in September 2000 at the Millennium Summit to provide targets against which progress in development could be measured and incorporated into the UN Millennium Project launched in 2002. The 'Biodiversity after Johannesburg' event examined the role of biodiversity in contributing to each MGD, for example MGD1, 'eradicate extreme poverty and hunger: poverty, hunger and biodiversity' and MGD7 'ensure environmental sustainability: water, sanitation, urban poverty and biodiversity', so that biodiversity could be fully incorporated into the UN Millennium Project. The Sixth Conference of Parties adopted '2010 – The Global Biodiversity Challenge' to develop effective implementation of the MDGs by 2010, essentially the means to fulfil the obligations of the Johannesburg Plan of Implementation.

In addition to the Rio CBD the Conference of Parties has adopted a second Convention known as the Cartagena Protocol on Biosafety, in January 2000. This is essentially a protocol to protect against the risks from modified organisms now being widely produced by contemporary biotechnology. In particular the protocol emphasised the need for countries to be able to make informed decisions before allowing **introduction** or movement of modified organisms across borders, which required a Biosafety Clearing-House to provide information, especially for those countries lacking any significant biotechnology expertise.

Those of you still awake will realise why the book *Biodiversity II* emphasised the rise of institutional structures for conserving biodiversity. Thankfully not all the CBD secretariat's work is so pitched towards large-scale bureaucracy. The CBD also promotes the International Day for Biological Diversity, on 22 May. The day is used to emphasise the importance of biodiversity, so, in 2004 the theme was 'Biodiversity: Food, Water and Health for All' and, despite all the problems and pessimism, report on progress towards fulfilling the goals of the CBD. On the 2004 day the UN was able to report that the CBD-COP had adopted the indicators required of signatories under the Johannesburg Plan of Implementation. You can find out how (or, in many cases if) your country celebrated the International Day for Biodiversity by visiting the CBD website.

The significance of biodiversity for human health and well-being is not confined to the developing world. Developed world governments have acknowledged the benefits from biodiversity. In the United Kingdom, central government has encouraged the incorporation of biodiversity within local economic and social development. For example the UK Biodiversity Action Partnership and the government Department for the Environment, Farming and Rural Affairs have developed the *Life Support* strategy which highlights how biodiversity benefits and stimulates community action, health, local economies, education and quality of life. Local Strategic Partnerships (a form of regional economic and social development) are required to incorporate these lessons into their planning. The UK government uses biodiversity, in the form of an index of abundance of bird species, as one of its five headline indicators of sustainable development in the United Kingdom, with the condition of Sites of Special Scientific Interest and the progress of Biodiversity Action Plans amongst an additional seventeen other indicators of progress towards sustainable development. The index starts from a baseline of 1970 and distinguishes different types of birds, such as woodland versus farmland, both of which have shown marked declines since the late 1970s. Presentation of the index includes Public Service Agreement targets, for example to reverse the decline of farmland birds by 2020, an example of the measurable indicators required by the UN led Plan of Implementation arising from Johannesburg.

At a local level in the UK, biodiversity has been taken up as a important asset around which health, employability, social inclusion strengthening communities can be developed. This is most obvious in urban environments. A fine example is the work of the Sheffield Wildlife Trust. British Wildlife Trusts have traditionally been a voluntary network focused on species and sites. The Sheffield Trust has developed much wider goals, using classic conservation projects such as tree planting, composting and urban farms to focus on health, training and employment, so that the benefits to city dwellers, many from poor and excluded backgrounds, become the main prize. Individual sites can offer a similar resource. In the heart of a deprived area of Newcastle upon Tyne, in north-east England, the Scotswood Natural Community Garden, set up in 1995 as a permaculture and wildlife garden, now hosts a wealth of activities all aimed at strengthening local communities, improving health, training, even down the raw pleasure of hosting parties and picnics (see Plate 2). These projects recognise the value of biodiversity as underpinning sustainability just as much as work in the tropical rainforests or savannahs.

Plate 2 Scotswood Community, Garden, Newcastle upon Tyne. This urban wildlife site combines many aspects of biodiversity from the obvious, e.g. wildlife and food production, to human health, well-being and social inclusion. These characters are crawling though a woven willow dragon's stomach.

The world discovers biodiversity

Definitions of biodiversity crystallised in the scientific and technical literature throughout the early 1980s (see Figure 1.1). Early discussion biodiversity embraced social, economic and cultural horizons. The concept of biodiversity was attuned with a popular mood. The imagery of nature's richness, imminent threat and human dependence on the continued welfare of the natural world was embraced rapidly and globally by the public, whilst providing a formal framework integrating emerging research themes for environmentalists.

The speed with which awareness of the term spread and intuitive response of a wide audience reflects a coincidence of environmental concerns and campaigns during the 1980s. Four mutually reinforcing trends came to a head by the late 1980s:

- activist-led environmental campaigns
- public awareness of wildlife via the media
- global-scale ecosystem degradation
- biotechnological interest in new products.

Activist-led environmental campaigns

Environmental campaigning organisations have blossomed since the mid-twentieth century. **WWF** (formerly the World Wide Fund for Nature which started out as the World Wildlife Fund) is typical of post-war international conservation bodies, combining research, funding conservation projects, political lobbying and public awareness. From the start WWF had a clear publicity agenda and was launched with a six-page, shock-horror

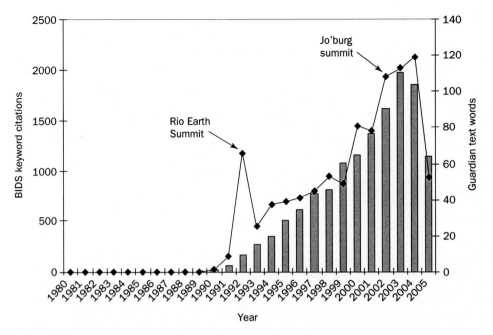

Figure 1.1 The appearance and increasing usage of the term biodiversity. Frequency in the scientific literature is shown by the numbers of articles containing the word 'biodiversity' in their title or abstracts, derived from the Science Citation Index *Web of Knowledge* (columns). Use in the mainstream media is measured by frequency of appearance of the word in articles in the *The Guardian* newspaper (line). In both cases 2005 data measures up to June

article in the British newspaper *Daily Mirror* linked to the plight of East African game, already seen as a tourist resource.

The 1980s saw the rise of headline-grabbing activism, most famously Greenpeace and Friends of the Earth, which both started from small protests in 1969. Greenpeace graduated from a small demonstration closing a US–Canadian border road in 1969, via a first sailing into a US nuclear test zone in 1971 to a string of publicity-grabbing actions. Swashbuckling physical bravery, often pitted against the media-unfriendly monoliths of government, big business and televisual baddies such as Soviet and Japanese whalers, earned huge public admiration (see Plate 3). The litany of success is remarkable. Incursions into French nuclear test zones in 1972–3 resulted in arrest. Smuggled film of French brutality to the prisoners discredited official denials of mistreatment and, by association, claims for the safety of tests. In 1975, the first anti-whaling mission resulted in film of harpoons being fired over the heads of Greenpeace inflatables into whales. In 1977, anti-seal-cull action in Newfoundland combined survival in atrocious Arctic weather with a visit by film star turned animal rights activist Brigitte Bardot ('blood, death and sex' – ideal). In the 1980s, Greenpeace turned to steeplejacking, using climbers to scale chimneys and factories to unfurl condemnatory banners. In 1983, a daring landing in Russia ended in a ten-hour boat chase towards Alaska. The Greenpeace boatman was captured but hidden film was later retrieved.

In the developing world campaigners fought a combination of government and business, at best corrupt and indifferent, at worst murderous. Tropical rainforests became a potent symbol (see Plate 4). In Brazil and South East Asia the fight against wanton clearance, often allied to rights of indigenous peoples, attracted global publicity. In Brazil encroachment of settlements and forest clearance, often illegal and enforced by private armies, had been highlighted since 1980 by the National Rubber Tappers Council, which

Plate 3 Greenpeace anti-whaling protest. Physical bravery and brilliant imagery caught the public imagination, reinforcing a growing concern for the diversity of life on earth. Photograph Greenpeace/Stone

Plate 4 Intact tropical rainforest. The photogenic quality of rainforests was an instrumental factor in the rise of environmentalism in the 1980s. This forest is on Mount Kupe, Cameroon.

was formed to defend the rights of all those making a sustainable living from forest resources. Headed by the charismatic Chico Mendes, the council's campaign culminated in a 1987 stand against eviction from an estate recently purchased by ranchers. Despite death threats and killings the Tappers held their ground. In December 1988 Mendes was shot dead by men he himself had credited as organisers of local death squads. By 1992 and the looming Rio Summit, their trial was still dragging on with an increasing sense that the system would save them. In the mean time Brazil was shocked by the international outcry and the first, albeit shaky, national environmental agency was set up. From 1986 tropical forest campaigns had been bolstered by international umbrella organisations (even if split between anti-trade, direct action versus dialogue, co-operation approaches). The rainforest campaigns brought together so many biodiversity themes: genetic, species and ecosystem loss, the importance of ecosystems for local need and global health, the role of indigenous people, undervaluing of natural resources, the potential uses of medicinal and food plants reflected in local knowledge.

Environmental campaigns remain a constant backdrop to biodiversity. In the developed world new battlegrounds have opened, such as direct action to destroy genetically modified crops planted in the United Kingdom. Other campaigns continue, e.g. whaling, whilst issues once regarded as closed have reopened, e.g. the resumption of seal culls in Canada in 2004. In the developing world the fight is much the same as ever, often pitching local campaigners who combine conservation of biodiversity with a concern for human rights against powerful vested interests and corrupt governments. Results can still be lethal. In Thailand in 2004 Charoen Wadaksorn, an environmentalist, was murdered after giving evidence to a parliamentary inquiry investigating corruption linked to building development, whilst in Ecuador in 2003 Angel Shingre was assassinated after publicising environmental damage caused by oil exploration in the Ecuadorian Amazon. International involvement may not offer protection. In the Philippines a project funded by the Frankfurt Zoological Society, the Philippine Endemic Species Conservation Project, has been targeted for its work in anti-hunting and anti-logging campaigns, and is forced to work from secure bases rather than stay in the forest.

Public awareness of wildlife via the media

Rainforests are deeply photogenic. Public interest in wildlife, especially the exotic, has an ancient history from amphitheatre spectacle, royal gifts, menageries and zoos. The appetite for natural history television appears insatiable, reinforced by high quality magazines such as *National Geographic* in the United States, *Animals*, relaunched as *BBC Wildlife*, in the United Kingdom and *Geo*, in Australia. *National Geographic* proved particularly influential with its photo essays and films of the great apes researchers, for example Jane Goodall and the Gombe chimps in 1963, Dian Fossey and the mountain gorillas of the Virunga Mountains in 1970. The images of heroic women in the wilds of Africa and the charismatic apes with whom they had made friends enchanted western audiences. However, these inspiring images hid many of the tensions and dilemmas of conserving wildlife in the developing world. Dian Fossey was mytholigised in the 1988 Hollywood film *Gorillas in the Mist*, which played on long-standing film traditions of white women in dark jungles, a touch of Tarzan, a hint of King Kong. But Fossey had been murdered three years previously, the culmination of a long and bitter struggle with local people, many of whom resented their exclusion from the nature reserves. The jungles of the films and photographs gave a very narrow view and western audiences could not appreciate the challenges facing both wildlife and people in Africa.

Wildlife media developed rapidly following the Second World War. The famous BBC Natural History Unit was founded in 1957 on the back of increasing success providing a non-sentimental, expert coverage of the natural world. By the late 1980s the developed world had a culture of wildlife television with personalities such as Jacques Cousteau and David Attenborough (see Box 1). Attenborough's *Life on Earth* series became a touchstone for the richness of life on Earth. Most famously his romp on the forest floor with a family of wild gorillas left a strong sense of kinship and threat. A massive, fascinated and knowledgeable audience has been able to see the literal diversity of life as never before. A night seldom goes by without TV film of animals eating each other, or mating, or sometimes doing both at once. The media impact can also be a source of problems. Superb documentaries have been accused of creating an image of a world still overflowing with life and a taste for the exotic so that the more familiar local animals and plants are ignored. The media have also been a major factor creating a sense of global anxiety. We are all instantly aware of threats to planetary health.

Box 1

The nature of televison

Natural history television has become a major genre in the developed world. Britain has led the way, from the early documentaries of Armand and Michaela Denis and Peter Scott to the globally distributed David Attenborough blockbusters or computer-generated imagery (CGI) animations of prehistoric life in the BBC's *Walking with Dinosaurs* or *Swimming with Sea Monsters* series. Modern audiences are shown more species than Victorian biologists could hope to see. However, whilst these programmes have un-doubtedly promoted the cause of conservation, they might also create problems. The style of programmes is very repetitive, offering a conveyor belt of one species after another, each showing off its remarkable adaptations. Many have storylines that show misfortune afflicting individual animals (a wildebeest eaten by lions here, a baby flamingo dead in a drought there) but almost always end with nature recovering and progressing serenely into the future. Both these patterns reinforce the idea of the balance of nature, the species well adapted, the future predictable and unchanging, nature as an idyll blown off course only by human interventions. The actual impression of a place can be misleading. The BBC's *Big Cat Diary* was a hugely successful weekly bulletin featuring the lives of lions, cheetahs and leopards in the Masai Mara game reserve. The footage showed endless plains and wild rivers, reinforcing the myth of wild Africa, but the game reserve is a comparatively small tract of land surrounded by farms and villages. The presentation of programmes has also been criticised for latent social and political messages. Many have a *Boy's Own* adventure style, the heroic presenter in faraway places following in a old tradition of western explorers braving danger to reveal the world. Local people are often invisible or portrayed as the cause of problems. The environmental writer George Monbiot went so far as to describe David Attenborough's programmes as representing 'The Planet of the Fakes'.

Global-scale ecosystem degradation

In the late 1980s the public, so supportive of green activists and familiar with wildlife, became aware of threats to planetary health that brought a deeper understanding than the familiar, grim but often rather local problems from oil spills (see Plate 5). Acid rain, ozone depletion and global warming combine similar themes. Each are complex and diverse

Plate 5 Human impact on rainforest. Logged forest but regenerating, South America.

processes but at their core is pollution of the atmosphere by human activity. In each case the chemical degradation provokes cascading complications. Acidic precipitation affects soils, forest health and freshwater life with economic losses to wood and fisheries. Human health is also at risk from increased dissolved metals and nitrogen compounds in water. Ozone depletion resulted in increased ultraviolet radiation reaching the earth's surface, potentially damaging to most life and known to suppress marine phytoplankton production with possible losses to fisheries. Human health is at risk from increased skin cancers. Global warming (still much debated but increasingly accepted) promises a host of impacts. Global shifts in climatic zones with associated shifts in natural and agricultural productivity, spread of tropical disease vectors, extremes of weather, sea level rises. Acidification, ozone loss and warming all suggest that damage to natural systems is not only a symptom but provokes further problems. All three are international in scope, encouraging a sense of global community and responsibility. For all their complexity the catchy titles were readily accepted. These problems combined ecology science, economic threat and environmental politics, just as biodiversity.

Biotechnological interest in new products

By the late 1980s there was an increasing awareness of the potential of biological resources, many yet to be discovered. The value of tropical forest species (mainly plants but also animals, fungi and microbes) for medical and food uses became an important argument used to defend forests. Some examples were simply known forest products that had always been used locally but could now be marketed effectively pandering to adventurous tastes in the developed world. Suggestions included the Lulo (*Solanum quitoense*), a fruit from Colombia and Ecuador especially good as a juice, and Amaranth (*Amaranthus* spp.), a protein-rich crop once used by the Aztecs but suppressed by Spanish colonists due to religious symbolism. This is now marketed as a health food in the United States.

Plate 6 Knock-on effect of rainforest logging: logged rainforest replanted with cannabis, South America. Pressures from the developed world are often cited as a cause of environmental destruction in developing countries. Here part of the damage is exported back.

Improvement of existing crops through discovery of new species or strains has been promoted through several famous examples. A fourth species of Maize (*Zea* spp.) was discovered by chance in Mexico in the late 1970s. Maize is one of the world's five most important crops. In 1972 botanists in Peru chanced upon a weedy new tomato. Seeds were collected but overlooked for several years. Genes from this species boosted soluble solids (the tasty component) of tomatoes by up to 50 per cent, potentially benefiting the US tomato industry by millions of dollars. Wild strains of crops may enhance resistance to disease, pests and climatic variation (see Plate 6).

Why biodiversity matters

The conceptual origins of biodiversity, rooted in reviews of the state of the global environment, show that the subject has always been more than an academic puzzle. Ecology, economics and environmentalism each brought an awareness of the significance of biodiversity. Expert and popular opinion contributed to a sense that biodiversity mattered and therefore so did our wise stewardship of the planet. Even opposing views, e.g. the clash between developing countries keen to utilise biodiversity and the developed world insistent upon its conservation merely highlighted this. The importance of biodiversity was explicit in Norse and McManus' (1980) paper with a review section entitled 'Material benefits of biological diversity' covering not only food, energy, medicine, raw materials, genetic discoveries but also ideas of human empathy with wildlife and philosophical attitudes.

The greatest insight, something humans had always sensed but made explicit only since the 1980s, is the importance of **ecosystem function** for planetary health. Ecosystems are made up of their living inhabitants and physical components. The activities of species drive processes, called **ecosystem services**, such as water gas cycling and purification, formation of soils, growth of food, fuels and products. Not only are these services vital to humans, but also they are free. An ecosystem function is the sum total of its services. Ecosystem functions depends on the organisation of the ecosystem, the diversity of taxa, their abilities and activity and distribution. Individual services are particularly dependent on species that have unique abilities (e.g. fixing nitrogen) or play a key role in processes without necessarily any special characteristic (e.g. keystone predators). Taxa with similar roles and abilities have been dubbed functional groups, acting as drivers of the ecosystem processes. These are so new and poorly understood that it is easy to draw the wrong conclusion, that if only we knew which the important driver functional groups and species were we could afford to lose the rest. The real lesson is exactly the opposite. We do not know and there is emerging evidence that ecosystem functions and their ability to recover from degradation improve with increasing diversity of species. All biodiversity is important. Combine the value of biodiversity that we use directly (e.g. crops) with the ecosystem services that biodiversity drives (e.g. oxygen production) and you arrive at figures such as the UN's US$33 trillion valuation. Even if we had the technology to replace all the biodiversity (which we don't) we do have to have the money.

In addition the importance of biodiversity touches all areas of our lives. Table 1.2 summarises specific historic, current and potential examples.

Not only does biodiversity exist as a resource we currently use or can foresee a use for, but also biodiversity is an insurance. We are increasingly aware of how much we do not know. Our stewardship of biodiversity is vitally important not just to us but also for the future. The concept of biodiversity may have a short history but the history of life on Earth teaches a telling lesson in uncertainty and of serendipity.

The next section explores the history of life and lessons from the past.

Biodiversity through time

The importance of time

Our awareness of current threats to biodiversity and prospects for future survival arises in part because we know of animals and plants that have become extinct in the past. News media feed an appetite for tales of dinosaurs, mammoths, living fossils and missing links. The biodiversity of Earth is not a static phenomenon. Processes that work over huge time scales (hundreds and thousands of millions of years, called **deep time**) have spawned present-day species and ecosystems that are conspicuously different from those of the past, yet life nowadays is intimately linked to the past by evolutionary lineages. Everything has an ancestor. Present-day species may be different but many lifestyles they lead are recognisably similar to those of extinct creatures. Many habitats are similar. Some were changed and some were dead but overall the diversity of life around us now is simply the continuation of the biodiversity of the past. Not that the inhabitants of Earth were just passive victims, tossed around by whatever the planet and solar system could inflict upon them. Life has survived on Earth for nearly 4 billion years. This may be more than just luck. Life itself seems to make the planet more habitable, buffering against many of the threats and adding to the resilience of the whole planetary environment.

To understand life nowadays, we need to know its history. To conserve biodiversity into the future we need to check for any lessons from the past. We need to know if extinctions now are qualitatively different from natural losses throughout the history of

Table 1.2 Historic, current and foreseeable importance of biodiversity to humans

	Historic	Current	Foreseeable
Ecosystem services	Pollination of crop plants by bees	Wetlands to clean pollution; effective as buffers against many pollutants, especially nutrient over-enrichment	Soft coastal engineering against sea-level rises from global warming
Medicine	Penicillin, antibiotic derived from fungus	Horseshoe crab blood used in bioassays of toxins	Rainforest plants; new drugs, to combat existing and new diseases
Biotechnology	Rattans, Asian forest palms used for fibres for many buildings and artefacts	Bacterial genes introduced to crops to confer resistance to insects	Metal-digesting bacteria; potential use to clean contamination and extract valuable metals
Environmental monitoring	Losses of bird species as evidence of DDT impact	European forests damaged by acid precipitation	Species range and habitat changes in response to global warming
Food	Irish potato famine due to over-reliance on one vulnerable cultivar	The weedy tomato, wild gene improved domestic cultivars	Several high value exotic rainforest fruits
Recreation	Victorian Fern craze, widespread hobby	Bird watching as a major recreational activity in developed world	Ecotourism; growth area for global tourism
Pets and domestic animals	Exotic zoo animals as gifts and status symbols	Value of current pet trade	Potential from use of historic breeds as we rediscover useful attributes
Political and social	Symbolic function in Celtic religion; animals as companions to St Cuthbert	Eco activists, e.g. Greenpeace and Earth First; highly visible political impact	Local ownership and control of resources in developing world
Security and resilience	Wild places as refuges in time of conflict warfare, e.g. forests of Eastern Arc Mountains, Tanzania	Control of biodiversity rich sites as a resource by opposing groups in conflicts	Healthy, functional ecosystems for post-conflict redevelopment

life, comparing the types of diversity currently at risk versus those wiped out in the past. We need to know if current losses are quantitatively different, if more species are being lost, at a faster or slower rate, over a larger or smaller area. The history of biodiversity provides insights, warnings and reassurance.

The history of life

How long has life existed on Earth?

Planet Earth is about 4.55 billion years old. A sense for this achingly long time grew only slowly throughout the nineteenth century as geological and biological inquiry unravelled the pattern and processes of the planet's natural environment. The 4.55 billion figure derives from the decay of radioactive forms of elements in rocks. The decay rates of **isotopes** (slightly different forms of individual elements) are precise. The time required to produce the ratios of different isotopes found in rocks nowadays can be calculated, establishing how long these rocks have existed. Moon and meteorite rocks created at the same time as planet Earth both predict the 4.55 billion age of the solar system (see Figure 1.2). Discoveries since the 1960s have cast a startling new light on the history and impact of biodiversity in deep time.

Era	Period titles		Million years ago ($\times 10^6$)
CENOZOIC	Quaternary	Holocene	
		Pleistocene	2
	Tertiary (T)	Pliocene	
		Miocene	
		Oligocene	
		Eocene	
		Paleocene	
			66
MESOZOIC	Cretaceous	K	144
	Jurassic	J	208
	Triassic	Tr	245
PALAEOZOIC	Permian	P	286
	Carboniferous	C	360
	Devonian	D	408
	Silurian	S	438
	Ordovician	O	505
	Cambrian	C	570
PRE-CAMBRIAN (pC)	Proterozoic		
			2500
	Archean		
			4500

Figure 1.2 The geological calendar. Note the time spans of the periods are not drawn to scale

The oldest rocks left to us are 3.8 billion years old from Greenland with other remnants between 3.6 billion and 3.36 billion years in South Africa and Australia. In 1977 a palaeontologist took samples from a South African remnant, 3.4 billion years old. The hard, flint-like rock was called a chertz, formed by mineral-laden volcanic water poured over thick mud. Microscopic examination of very thin sections of the chertz showed minute round, occasionally dumbbell-shaped structures. From their size, shape and apparent division, they were eerily like modern-day bacteria to look at. This discovery prompted searches of other ancient rocks, most famously Fig Tree formations (South Africa, 3.36 billion), Gunflint (Canada, 2.5 billion) and Bitter Springs (Australia, 1.1 billion). All contained **microfossils** of single celled life, with increasing diversity in younger rocks. The Gunflint samples yielded spheres, rods, filaments, stalked blobs and tentacled forms. The very oldest rocks known do not harbour microfossils. However, the ratio of carbon isotopes, C12 and C13, in the rocks is unusual, with C12 higher than expected if **abiotic** (non-living) inorganic chemistry processes were solely responsible. Elevated C12 may be another sign of life due to preferential use of this isotope by **photosynthetic** organisms.

The overriding message of these rocks is that life goes back through deep time as far as we can delve. From perhaps 3.8 billion years ago Earth has supported life. This early origin and continued survival in one form or another suggest that Earth is very conducive to life. The planet is near enough to an external energy source, the Sun, to prevent freezing, but not too close that elements evaporate and the surface is destroyed. Earth is not so small that the atmosphere is stripped away, not so large that the atmosphere is dense and muffling. The conditions allow carbon-based chemistry to flourish. Carbon is abundant and reactive, without being dangerous, and the basis for the most complex and varied chemistry. Condensed water is plentiful. Water is the best solute for most chemistry and is liquid across a range of temperatures, fast enough to allow chemical reactions but not so fast that systems collapse. In addition water's surface tension and heat exchange properties have been exploited by life.

The early origin of life and continued survival despite terrible natural disasters is profoundly reassuring. Life is robust and resilient. Although humans have wrecked havoc with some habitats, nothing we have done has threatened the overall survival of life. Other lessons are more challenging. For 3.1 billion years, 80 per cent of the history of life, biodiversity was dominated by bacteria, **cyanobacteria** and other life forms now dismissed as microbes, as if size alone was the criteria for judging the importance of biodiversity. Judged by their monopoly of deep time, the biodiversity of Earth is the story of the bacteria.

Biodiversity rising

The history of biodiversity is a epic adventure. Throughout there are patterns and crises that have a particular relevance for our understanding of the patterns and crises nowadays.

The age of bacteria

From their ancient origins to 670 million years ago biodiversity was dominated by bacteria and their kin, a **kingdom** of life called the **prokaryotes**. Prokaryote evolution is the foundation for the diversity of life, genetic, metabolic and ecological, today. Their role in Earth's history shows the power of living things to alter the environment at a planetary scale.

The precise environmental chemistry of the **Archean** is disputed but general charac-teristics were an atmosphere of nitrogen, hydrogen, carbon dioxide, water vapour and traces of other gases with the exception of oxygen. The lack of oxygen affected ancient seas, creating reduced conditions which determine the forms of chemical ions present. Earth's atmosphere is now predominantly nitrogen with about 20 per cent oxygen. This archaic environment provided the cradle in which prokaryotes developed a diversity of metabolic systems for three purposes: releasing energy, building carbon reserves and tolerating extreme environments. This variety of metabolisms and tolerances still exists and at a fundamental level represents a much greater diversity than the difference between a jellyfish, an ant and a human, all of which share the same energy-processing metabolic system.

The Archean spawned microbes able to survive in a range of frighteningly severe habitats, viciously alkaline soda lakes, thermal vents discharging water hotter than 100°C and desiccating salt pans. Many of these especially tolerant bacteria belong to a group called the **Archeabacteria**, which are separated from the other bacteria, called the **Eubacteria**, the two **domains** combining to form the prokaryotes. All other life belongs to a third domain, the **Eukarya**, often called the **eukaryotes**. These three domains are distinguished by the biochemistry of structures inside their cells called **ribosomes**, essentially sites where genetic code is read and used to build protein. This three-domain divide, Archeabacteria, Eubacteria and Eukarya, is the fundamental division of bio-diversity on Earth.

The diversity of bacterial metabolism reflects the lack of gaseous oxygen in the Archean atmosphere. Free oxygen is toxic to many bacteria, their metabolic chemistry being disrupted by this highly reactive chemical. However, free oxygen's origins are also from life, an example of biodiversity changing global environments and the first and most severe biodiversity crisis to afflict Earth.

Oxygen is the waste product from **photosynthesis** splitting water to obtain hydrogen ions. Cyanobacteria are credited with this atmospheric revolution. Fossil evidence for Cyanobacteria goes back to 2.6 billion. From then until about 1.8 billion years ago, oxygen production was mopped up, reacting with abundant ions such as calcium and iron in the sea. Thick Iron Band deposits some 2.0 billion years old are evidence of particularly rich deposition. As iron and other chemical sinks for the oxygen were used up, atmos-pheric oxygen increased. The replacement of a carbon dioxide–methane rich greenhouse atmosphere with oxygen may have sparked the first known Ice Age. For many prokaryotes oxygen was a deadly poison. Global ecosystems were destroyed. Remnant communities were banished to habitats beyond oxygen's reach such as waterlogged mud and the deep sea. For others this was an opportunity. Respiration using oxygen to release energy (**aerobic**) is more efficient than **anaerobic**. Prokaryotes able to use oxygen thrived. In the wake of the oxygen holocaust the third domain, Eukarya, appeared, the result of an extraordinary biodiversity link up.

Eukaryotes differ from prokaryotes by possession of discrete, membrane bound structures inside their cells, called **organelles**. Different organelles specialise in different tasks. Two of the most distinctive are **mitochondria**, the site of aerobic respiration, and **chloroplasts**, the site for conversion of sunlight radiation to stored chemical energy. Both chloroplasts and mitochondria have their own genetic information, separate from that of the rest of the cell. When the cell divides, the mitochondria and chloroplasts duplicate themselves simultaneously but separately. These two organelles are very like stripped down bacteria, retaining their specialist metabolic skills but harboured in the bigger cell. Mitochondria and chloroplasts appear to be remnants of once independent bacteria, subsumed into a third bacterium, the main body cell. In addition the motile hair-like cilia outside many animal cells are uniquely similar to the structure of motile spirobacteria. There are still hints of this origin of eukaryotic cells, termed endosymbiosis, amongst modern creatures. One of the most extraordinary arrangements is *Mixotricha paradoxa*, superficially a single celled animal found in the hind guts of termites but actually a collection of at least five types of pro and eukaryote (see Box 2). In biodiversity terms eukaryote cells, yours and mine, are an intimate synergy of different bacteria, individual

Box 2

Mixotricha paradoxa: a creature or community?

Mixotricha paradoxa (the Latin translates as the mixed hair paradox) lives in the hind guts of an Australian termite, *Mastotermes darwinenisis* (see Figure 1.3). Many termites rely on an exotic gut fauna to digest wood, the flora including many strange prokaryotes and single-celled eukaryotes. *Mixotricha* appeared to be just another gut ciliate but detailed analysis of the structure reveals a multiple symbiosis. The pear-shaped body is a eukaryote cell, a Trichomonad protistan. Inside the cell are bacteria which may function in place of mitochondria, which the creature lacks. On the surface are three types of spirochaete bacteria, the most numerous (500,000) undulate to move the cell, a second bacterium which is part of the anchor mechanism for the small spirochaetes, and a few larger spirochaetes. Other examples of spirochaetes forming motility associations have been found amongst other termite gut flora.

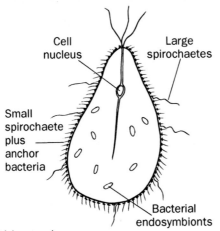

Figure 1.3 *Mixotricha paradoxa*

Source: Redrawn from Margulis 1993.

identities sunk into a co-operative venture, combining the diversity of metabolic systems. Every eukaryote is a bacterial homunculus.

So the fundamental biodiversity of life is three domains, Archeabacteria, Eubacteria and Eukarya. Prokaryote diversity is characterised by their varied metabolic skills. The origin of atmospheric oxygen was the result of prokaryote photosynthesis, even the smallest, most ancient life was capable of modifying the planetary environment. A global biodiversity catastrophe due to oxygen toxicity and climate change occurred but even this most terrible event did not wipe out life. Oxygen tolerant prokaryotes took over and new cellular organisms developed from an intricate association of prokaryotes.

A snowball's chance in Hell

Between 670 million and 550 million years ago, the Earth went through an extraordinary epoch, called the Vendian. Within this period the most serious global catastrophe since the advent of oxygen occurred but in its aftermath large animals appeared with startling speed, as if in part forged by the grim conditions of what has been named Snowball Earth.

Between 590 million and 570 million years ago Earth underwent an almost complete global glaciation. Geological strata contain huge glacial deposits from what were sea level, equatorial regions. Iron-rich deposits, suggesting a marked decline in atmospheric oxygen and reduced Carbon 12, a signature of life, both suggest a serious decline in biological activity. Oddly these signs of massive glaciation are overlain with carbonate rocks that require carbon dioxide rich, warm seas for deposition. The Snowball Earth hypothesis has been developed to explain these extraordinary features. The hypothesis starts with the position of continents at about 770 million years ago, most of the land masses being small, fragmented and equatorial. Climate models suggest that this arrangement would result in high rainfall and geochemical processes stripping carbon dioxide from the atmosphere. A 6 per cent weaker sun combined with increasing reflection of solar energy from the growing polar ice caps and perhaps variations in planetary tilt relative to the sun resulted in a runaway glaciation with one kilometre thick ice covering most of the Earth and surface temperatures of −50°C. Only oceanic depths, heated from the Earth's core, and volcanic peaks did not freeze over. With biological systems closed down or banished to the ocean depths, carbon dioxide from volcanic sources began to increase. Some models suggest a thousandfold increase of carbon dioxide, precipitating a very rapid thaw, super greenhouse effect and a planet bathed in concentrated carbonic acid rain and now at +50°C, allowing the deposition of carbon-rich sediments over the glacial debris − 'Freeze-Fry', as the suggestion has been elegantly summarised.

End points for the global glaciation are estimated at between 580 million and 595 million years ago. By 575 million years ago substantial Ediacaran animals occupied the seas (Box 3). The speed of this diversification is remarkable. Proponents of Snowball Earth have suggested that evolutionary radiation of the deep sea survivors was both rapid and novel as they recolonised the emptied seas. It had been a close-run thing. Snowball Earth was one of perhaps two crises that could have destroyed life on Earth, as the ability of biodiversity to regulate the global environment was overwhelmed by the unusual combination of circumstances.

Lilos and hallucinations: nature experiments with animals.

Darwin fretted at the lack of Pre-Cambrian fossils. In 1947 on an Australian mountain range called **Ediacara**, formed of rocks some 570 million years old, a Pre-Cambrian fauna was discovered.

The Ediacaran fauna is now known from similarly aged rocks around the world, a global diversity of shallow marine habitat animals but one that does little to explain the Cambrian explosion. The Ediacaran ecosystem is represented by beautifully preserved fossils of entirely soft-bodied animals. If animals they be, most resemble frisbees, ferns and rings. There are no hard structures, no mouths, no sense organs, no limbs. The flat discs or lobes are segmented, much like an inflatable bed. The Ediacaran animals look like a natural experiment into life as a lilo (Box 3). Initial interpretations placed the fossils as precursors of known animal lineages such as worms or anemones. An alternative is that the Ediacarans are an entirely separate lineage, an experiment with no obvious features to justify links to later groups, a novel ecology. Their flat and fronded shapes may have relied on the large surface area relative to volume to obtain nourishment and exchange gases, rather than internal organs. Quite what they are remains a source of debate. They are animals utterly alien in design.

As rapidly as they appeared and dominated so the Ediacaran fauna was lost. The Pre-Cambrian–Cambrian boundary is uncertain. Between the Ediacarans and the Cambrian explosion, around 530 million year ago, lie the so-called 'small shelly fauna' − odd scales, spines and tubes, perhaps the skeletal structures once attached to softer bodies or perhaps complete cases and bodies of another strange experiment. Ediacara, the small shelly fauna

Box 3

The Garden of Ediacara

The strange, plant like shapes of many Ediacaran animals gave rise to an image of the community as the Garden of Ediacara, an echo also of the Garden of Eden at the dawn of obvious life (see Figure 1.4). The oldest known animal fossils are Ediacarans between 565 million and 575 million years old in rocks in Newfoundland. These include substantial creatures, 2 metre long fronds. The fossils suggest a succession of species over differing depths, the richness and abundance being very similar to modern communities of deepening seabeds. There is also global variation in the distribution of types of Ediacarans, with some types restricted to what were equatorial waters. Whilst they may look unfamiliar, the variety and distribution of Ediacaran fossils suggest that they lived their lives to the same ecological rhythms that drive modern communities of animals.

Figure 1.4 Ediacaran animals

and Cambrian explosion may represent three quite separate developments, nature experimenting with how to be an animal, each dominant within the 40-million year span, but with only the Cambrian fauna leaving any descendants. The biodiversity of present-day animal life is the progeny of this third fauna, the Cambrian experiment.

The sudden diversity of animal life that had perplexed Darwin appeared to contain representatives of all the major animal lineages alive nowadays. Ever since the Cambrian period, animal diversity has been variations on the same themes. Although animal diversity may now seem extraordinary, the majority of species are developments of but one design, the insect, and mocked as '600 million years of fussing'. The image of an ever-increasing diversity of animal life with its latent messages of progression and improvement towards the pinnacle that is humanity has also been undermined by analyses of the early Cambrian faunas, suggesting that some fossils represent animal designs that have left no descendants. So, the history of animal diversity since the Cambrian may be one of fundamental loss of **phyla** but increasing species diversity within a handful of those remaining.

This sense of loss is epitomised by the Burgess Shale animals, an early Cambrian marine fauna from about 525 million years ago, exquisitely preserved in shale deposits in the Canadian Rockies. The Burgess Shale fauna appear to be whole communities apparently swept to their doom from shallow continental shelves by mud slides. Many Burgess Shale animals can be readily allotted to known groups. Other Burgess creatures have provoked fierce debate, often going to the core of how we think of the natural world; are they just unusual examples of familiar groups or wholly different animal designs? The American palaeontologist Steven Jay Gould championed the idea that the Burgess fauna included many fundamentally different animals designs compared to a more limited selection left on Earth nowadays. This revision of early biodiversity cast a darker shadow on our understanding of life. There are no obvious hints as to the likely survivors or the doomed, winners and losers. Survival of lineages may be a matter of luck, as much as any deterministic, predictable pattern. No animal seems a better example of this extraordinary fauna and status of weird wonder than *Hallucigenia*, described in Box 4. Conversely Conway Morris, who had worked extensively on the Burgess fossils, counter-attacked with evidence that most, perhaps all, of the animals had affinities to known phyla and he has subsequently developed arguments for the almost inevitable, uncontingent, evolutionary road leading towards sentient life on Earth. The precise status of 525 million year old shrimps may seem an arcane battleground, but the argument is really about fundamental views of nature; Gould's chancy, contingent, accident-prone vision can be interpreted as a Marxist analysis of the history of life, Conway Morris's gradualistic, inevitable, rise of sentience, with his talk of 'deep structures' underpinning evolution, presents a less revolutionary stance. Sadly Gould died in 2002. Meanwhile Conway Morris has been instrumental in describing 530 million years ago fossils from China that appear to show key affinities with Chordates. These strange Vetulicolians and Yunnanozoans have got him into new arguments about our representations of life on Earth.

These early faunas and their fates send important messages. Initial animal biodiversity suggests experimentation with biodiversities now lost. Different fauna waxed and waned. The fate of individual lineages may be unpredictable. Present-day animal diversity is but massive variation on a few themes; to mangle Henry Ford's famous maxim on car colour, you can have any animal you like so long as it is an insect. Animal diversity is no more than one outcome from a mass of permutations. No lineage has a guarantee. Nature is not sentimental to her first born.

Founding dynasties and terminal disasters

From the Cambrian to end of the Permian (the Palaeozoic, 345 million years long) global biodiversity increased in two major ways: first, the diversity of animal species, primarily in the seas; second the colonisation of land opening up new ecosystems.

Throughout the Palaeozoic marine faunal richness increased, measured as families or genera, even allowing for artefacts such as greater availability of younger rocks. Different

Box 4

Hallucigenia sparsa: weird wonder or upside-down worm?

In 1977 the journal *Palaeontology* contained an article describing an animal named *Hallucigenia* in honour of its baffling form. Pictured tip-toeing over the Cambrian ooze on stiletto spines, the description ended coyly, 'its affinities remain uncertain'. The article speculated on the lifestyle of this 2–3 cm long monstrosity, the concentration of individuals around remains of a worm only adding to the eerie quality: 'the spines could have been embedded in the decaying flesh'. In 1985 a review of the Burgess Shale had *Hallucigenia* listed under miscellaneous animals, by now perhaps 'grasping food with its tentacles'. With the publication of Stephen Gould's *Wonderful Life* (1989), *Hallucigenia* became an icon of all that seemed mysterious about the Cambrian fauna, of the 'stunning disparity and uniqueness of anatomy'. Yet Gould was also wary that so strange a beast may be reinterpreted. By 1991 published evidence of other worm-like fossils from the early Cambrian, with obvious legs but also bearing spines and plates, forced a revision of *Hallucigenia*. It is now generally regarded as allied to the phylum Onychophora, represented nowadays by velvet worms, mysterious inhabitants of the tropical rainforest floor. Gould could not help a pang of disappointment, writing, 'I regret the loss of a wonderful weirdo'.

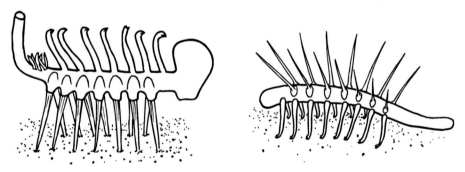

Figure 1.5 *Hallucigenia sparsa*; stilt walking alien from Conway Morris, 1977; reinterpreted as spikey worm from Ramskold and Xianguang, 1991

animals dominated at different times, important ecosystems were established and different ways of life became established, e.g. filter feeding by clams and brachiopods in the Ordovician. Arms races were won and lost as predator and prey locked into escalating spirals to overcome one another.

The late Ordovician witnessed the first land plants and invertebrate animals, the centipede/millipede/insect line, spiders, woodlice. Some of the earliest land plant communities, including the famous *Rhynia*, show damage that resembles attacks from herbivorous invertebrates. The establishment of land plants was based on an association with fungi, sharing nutrients, a co-operative mutualism that is still the foundation of terrestrial vegetation diversity. Land opened up new opportunities. The forest created habitat, physical space and new resources. The predominance of insect species diversity nowadays depends on groups intimately co-evolved with plants. Whilst these modern insects and the plants appear later, the terrestrial insect–plant link was established in the Palaeozoic (see Figures 1.6 and 1.7).

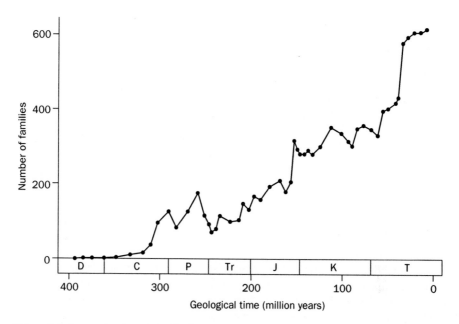

Figure 1.6 Increasing insect family diversity through time. The drop at the end of the Permian probably represents real decline, the rapid rises in the Tertiary (T) are artefacts of good preservation

Source: Redrawn from Heywood (1995).

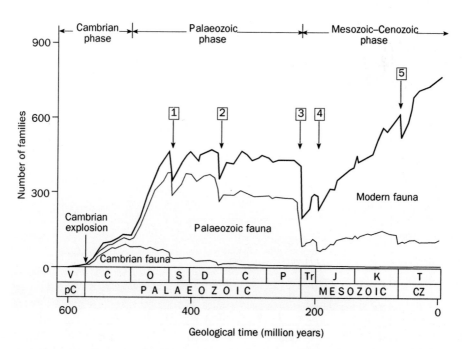

Figure 1.7 Animal family diversity since the Cambrian explosion. Diversity generally increases but different fauna dominate different periods and there are occasional mass extinctions. The five mass extinctions are marked: 1 Ordivician; 2 Devonian; 3 end Permian; 4 Triassic; 5 end Cretaceous (K/T)

Source: Redrawn from Heywood (1995).

The end of the Palaeozoic is marked by a global catastrophe, the Permian extinction and sudden, global disaster about 251 million years ago resulting in the loss of over 95 per cent of all species, perhaps threatening the very existence of life on Earth. This mass extinction coincides with a period of ferocious lava eruptions called flood-volcanism, which have created what we call the Siberian traps. Essentially this event was an out-pouring of lava lasting over a million years, covering 1.6 million square kilometres and dumping 2 million cubic kilometres of lava across the landscape. The global impact arises from huge additions of atmospheric greenhouses gases, provoking warming by +6°C. The oceans appear to have become widely anoxic, perhaps exacerbated by the arrangement of continents as they grouped together. Changes to levels of oxygen and carbon isotopes in the rocks suggest catastrophic breakdown of biogeochemical systems as biodiversity's ability to drive global geochemical systems was overwhelmed. Recovery from the Permian extinctions was the start of a contest between different groups of vertebrate to dominate terrestrial habitats, resolved (for the time being) 179 million years later at the end of the Mesozoic with the extinction of the dinosaurs.

During the early Triassic five dynasties of large animals vied for the land: labyrinthodont amphibians, three derived from reptilian stock (Rhyncosaurs, Therapsids, dinosaurs) and Synapsids ('mammal-like reptiles'). Each dominated at different times, but the sequence defies the prejudice inherent in classifications that hint at mammalian superiority. The ultimate winners were the dinosaurs, dominating the large terrestrial animal niches from the late Triassic to the end of the Cretaceous. Terrestrial biodiversity toward the end of the Mesozoic would have looked largely familiar to us: broad-leaved and coniferous woodlands, butterflies and birds flitting from flower to flower, and small mammals, frogs and toads in the undergrowth. Only the occasional dinosaur and absence of grass would be strikingly different (see Figure 1.8).

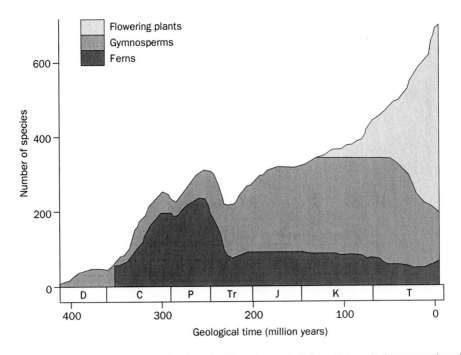

Figure 1.8 Vascular plant diversity since the Devonian period. Overall diversity increases though different groups dominate at different times

Source: Redrawn from Groombridge (1992).

At the end of the Cretaceous the dinosaurs and many even more ancient lineages were completely wiped out. In 1980 a massive meteorite impact was proposed as the cause by father and son team Luis and Walter Alvarez and colleagues Frank Asaro and Helen Michel. This idea was prompted by discovery of high concentrations of the element iridium in deposits at the Cretaceous–Tertiary or **K/T boundary** outside the town of Gubbio in Italy. Iridium is very rare on the Earth's surface but abundant in meteors (see Box 5). The elevated levels in the Italian samples suggested debris deposition following a meteor impact. The drama of this scenario provoked fierce debate, partly responsible for our increasing awareness of the extent and nature of extinctions throughout the history of life and comparison with current threats to biodiversity. The Chicxulub crater in Mexico is now widely acknowledged as the impact site, although there are other major craters from this period, such as Boltysh in Ukraine, suggesting that the Earth may have suffered a terrible coincidence of destruction. The sudden, catastrophic hypothesis is now widely accepted but perhaps the dinosaurs were doubly unlucky, caught in a double whammy of an impact and a major flood-volcanism event which created the Deccan traps in India. The Deccan traps volcanism may have released sufficient greenhouse gases to cause a 3–5°C rise in atmospheric temperatures. The volcanism started before the impact date, weakening systems before the impact overwhelmed them. One analysis, of planktonic foraminifera, has identified a grim series of events in the last 500,000 years before the K/T boundary; global warming of 3–4°C, major volcanic activity, and evidence of multiple impacts, the final one representing the end of this havoc. Other major impacts are not associated with mass extinctions, e.g. the Manicouagan in Northern Canada, estimated to be from a 100 km diameter asteroid dated at 214 million years ago, is not associated with a mass extinction. At Manicouagan, ground zero was a vast, arid Triassic desert with little to destroy and limited water and volatile geology to vapourise. The impact hypothesis has prompted a serious search for other likely asteroid or comet threats to Earth. The Spacewatch scheme in the United States estimates numbers of very dangerous 1 km+ size at between 1000 and 2000. Analysis should have identified any serious collision courses by 2008. Meanwhile in 1994 an asteroid passed by Earth at only 105,000 km.

The contest between vertebrate dynasties to dominate the large animal niches on land again shows the turnover of biodiversity. Extinctions and radiations, both natural processes, are equally important. The history of land vertebrates is one of shifting dominance. The Mesozoic world was generally warmer than nowadays, a greenhouse world benefiting the dinosaurs, versus the icehouse currently favouring mammals. If global climate changes markedly, the ecological balance of advantage may shift, though dinosaurs as such are long gone. The dinosaurs' fate has prompted some of the hardest revisions to our idea of extinction, once so firmly tied to ideas of inferiority and obsolescence. The danger may not be so much bad genes but rather bad luck.

How many species have there been?

Despite all the extinctions and losses, life has never been utterly destroyed. The diversity of life with which we share the planet can be reduced to a mundane equation: species diversity today = **speciation** – extinctions. What we have left is a surplus and, although species diversity has generally risen over time, the majority of species that have ever lived are now extinct. Estimates put the current animal and plant species diversity at 2–5 per cent of the historic total. Such calculations rely on historic evidence for species diversity and longevity, data most reliably gleaned for marine invertebrates that have the most complete fossil record. Individual species vary greatly in lifespan, with many fossil mammals lasting only 1 million years; the average for marine invertebrates is 11 million

---

Box 5

Evidence for the K/T impact hypothesis

There is now multiple evidence for the impact hypothesis.

Iridium Anomalously high levels of iridium have now been found at over 100 K/T boundary sites worldwide.

Shocked quartz Quartz grains with distinctive shock-induced stress patterns, even as far as molecular rearrangements, have been found only at known meteorite impact sites. These require extreme conditions to create and are found at K/T boundary sites.

Mass carbon deposits Unusually large deposits of charcoal and soot found at K/T boundary sites are suggestive of massive global fires ignited by the impact.

Spherules Spherical particles are formed by cooling of molten droplets. Droplets found in K/T boundary layers do not fit the chemical make-up and form of similar droplets that volcanoes may produce.

Diamonds Diamonds from K/T boundary sites have carbon isotope ratios similar to those from meteorites, not terrestrial sources.

Amino acids Analysis of amino acids from K/T boundary material showed similarity to those known from meteorites, not terrestrial sources.

Tidal wave debris Thick, jumbled deposits of material, typical of massive tidal waves, have been found, particularly around the Caribbean.

Chicxulub crater In 1990 the 200–300 km wide, 65 million year old Chicxulub crater at the north tip of the Yucatan peninsular was widely publicised as the impact site. In addition this massive crater lines up in an arc with several smaller craters scattered over the globe and dated to the same time, suggesting a multiple or glancing impact. Recent, and hotly contested, analysis of K/T sediments in Mexico has been used to suggest that Chicxulub was the earlier of at least two impacts, separated by a few hundred thousand years (see Figures 1.9, 1.10, 1.11 and 1.12).

---

years and for all species taken together 5–10 million years. If we combine estimates for the known existence of animal life, average duration of species, the numbers of animals alive now and adjustment for declining background extinction rates, one estimate of the numbers of animal species that have ever lived comes out at 18,375,000.

Natural extinctions

Types of extinction

The natural fate of all species is extinction. This can mean the total loss as a species dies out. Alternatively as a species evolves, individuals will eventually become so different from the ancestral species that it is effectively no more.

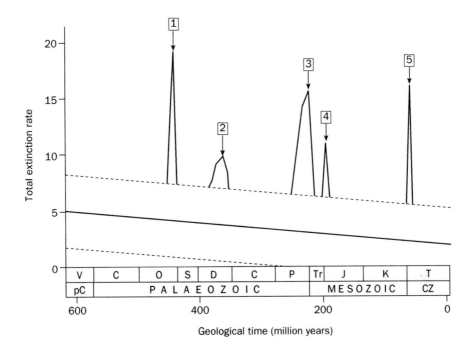

Figure 1.9 Extinction rates of marine animal families since the Cambrian period. Extinction rates are given as numbers of families lost per million years. The solid line shows the average rate for background extinctions, which all lie within the dotted boundaries. The five major mass extinctions are revealed by spikes of increased extinction rates

Source: Redrawn from Heywood (1995).

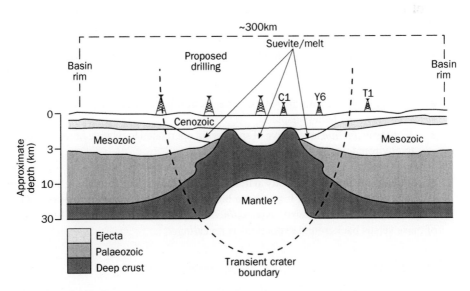

Figure 1.10 Simplified cross-section of the Chicxulub crater site, Mexico. The unusual rock structures were revealed by geological drill sites (C1, Y6, T1) used for oil exploration. The mantle, Palaeozoic and Mesozoic rocks are severely disrupted. Suevite and melt rocks are various products of catastrophic disturbance. Cenozoic rocks have subsequently reburied the site

Source: Redrawn from Smit, J. (1994) Blind tests and muddy waters, *Nature*, 368, 809–810.

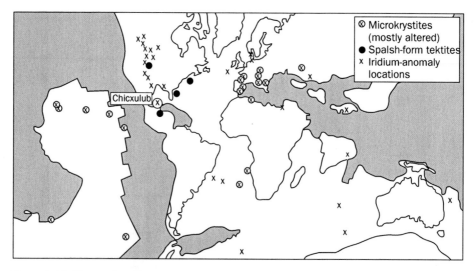

Figure 1.11 The continents at the end of the Cretaceous when a giant meteor hit. The Chicxulub crater site is marked along with sites from which fall-out debris such as high Iridium contamination, splash-form tektites (droplets of glass) and microkrystites (cristallised droplets of rock) have been found

Source: Redrawn from Smit, J. (1991) Where did it happen? *Nature*, 349, 461–462.

The lifespans of genera have been tracked through the fossil record. Most genera have comparatively short lifespans, some have longer, a few very long. Most genera share three characteristics; few species, few individuals in each species and small geographic ranges that together can account for short lifespans of species. All species encounter good and bad years. A very bad run may wipe out a species. In any set of species for every few doing very well, there will be some doing very badly, to the point of oblivion. This gradual attrition wipes out species by nothing more than a simple statistical probability. David Raup (1993) distinguishes three modes of extinction:

- *Field of bullets* describes random extinction unrelated to relative fitness of species: the name stems from an analogy of soldiers advancing across a battlefield being mown down at random.
- *Fair game* extinctions are selective losses of less fit species, the classical Darwinian survival of the better adapted.
- *Wanton extinction* is selective of species but not related to relative fitness or adaptation to environment.

A continual attrition is perfectly natural; some species lost to nothing more than the statistics of bad luck. Such losses have been labelled *background extinctions*. They are eclipsed by the sudden, global, *mass extinctions*, pivotal in the history of life. The nature of mass versus background extinctions is revealing.

Background versus mass extinctions.

From the mid-1970s a group of American scientists, notably John Sepkoski and David Raup, compiled a detailed compendium summarising known origin and extinction dates for families of marine animals. The work coincided with the revived interest in mass extinctions and catastrophes following the K/T meteor impact hypothesis. The database was a treasure trove for measures of variations in extinction rates, in particular comparison

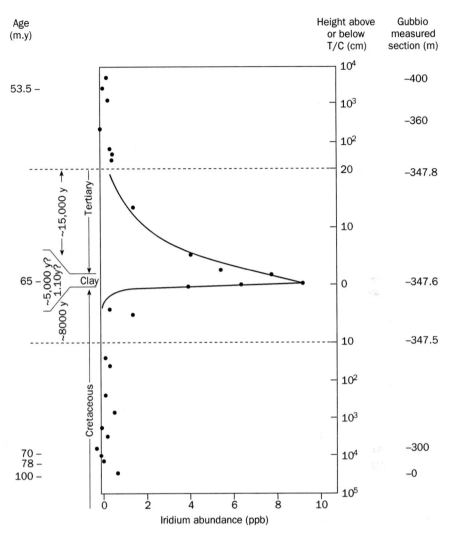

Figure 1.12 Simplified diagram of the 'Iridium anomaly spike' from Alavrez *et al.*'s 1980 paper proposing a meteor impact as cause of the K/T boundary extinctions. Iridium abundance is shown from rocks near Gubbio, Italy, before, at and after the KIT boundary along with rock depths and times

Source: Redrawn from Alvarez *et al.* (1980).

of background versus mass extinction rates. The work revealed some striking patterns, notably peaks of extinction every 26 million years with five especially serious extinction events between the Cambrian and K/T boundary. The periodicity inspired speculation of dramatic causes, mostly extraterrestrial, whether meteorites, super novae or an unknown death star dubbed Nemesis. More recent analysis of periodicities for fossil plankton which provide detailed sequences of data suggest that many groups show rhythms of between 26 million and 29 million years, but are tracking global environmental pacemakers such as sea level and climate, with volcanism and impacts creating additional crises.

Mass extinction rates appear catastrophic. There have been five major mass extinctions and analyses provide stark estimates of species losses. In percentage terms estimates are Late Ordovician 84–85 per cent, Late Devonian 79–83 per cent, End Permian 95 per cent, End Triassic 79–80 per cent and the K/T event 70–76 per cent. There is some evidence of flood-basalt volcanism associated with all except the Triassic. Perhaps the most serious events require a double-whammy: the Permian greenhouse/anoxia, the Cretaceous impact and flood-basalt. Nonetheless most extinctions have happened outside of these catastrophes, the result of background extinctions. The fossil record suggests an average species lifespan of about 5 million years, for any one species equivalent to 20 per cent chance of extinction every 1 million years. Expressed as the probability of a species going extinct per year this is 0.0000002. Assume current estimates of species diversity on Earth now are correct at between 10 million and 30 million, this background extinction rate predicts yearly losses of between two and six species. Background extinction rates show signs of a steady decline over time for some taxa, notably marine invertebrates, but perhaps increases for others such as land plants. A declining rate may reflect the resilience of surviving taxa, winnowed by previous losses. Extinction rates vary with taxa. Differences of lifespan between taxa may result from their biological traits. These attributes may be the now familiar factors of range and abundance but other characteristics can be important. Analysis of extinction rates for marine Gastropod molluscs (snails and their kin) shows that the feeding ecology of the larvae is a better predictor of susceptibility to extinction than adult feeding. Taxa characteristic of ecosystems with a history of vulnerability such as coral reefs may always be at risk.

To separate background from mass extinctions may be an artificial divide of a seamless gradient of varying rates. Mass extinctions affect many more species, impact globally and are, geologically, short events but these are simply increases in rate and scale rather than qualitatively different from background losses. Mass extinctions may be nothing more than the extreme end of a continuum, the very rare end of the skewed distribution. Mass extinctions are better defined by their impact than quantities such as area afflicted, numbers lost or time span (Box 6).

Box 6

Characteristics of mass extinctions

Qualitative losses Major lineages are entirely wiped out. Several are wiped out simultaneously. Many habitats are affected, both marine and terrestrial. Disparate taxa are affected. Ecological stage is reset via ecological collapse as dominant taxa are replaced.

Scale Global impacts are taxonomically broad and rapid.

Survivorship Luck? Survival of taxa may be pure good fortune or due to a characteristic that turns out to be fortuitously useful but could not be a prescient adaptation to impending danger. Such adaptations, beneficial in retrospect, are called **exaptations**.

Luck continued Taxa that survived previous mass extinctions survive subsequent events.

Range It is useful to have the ability to adjust geographic range rapidly or to be widespread already.

Tolerance The ecology of taxa may make them largely immune to a particular event. Freshwater and deep-sea fish survived the Devonian mass extinction that hit shallow marine taxa hard.

Mass extinction events coincide with major changes to the global environment but no one type of event can be matched to all mass extinctions. Four of the five extinctions between the Cambrian and K/T boundary have been linked to flood-volcanism and at least two with impact events. The Late Devonian mass extinction comprised losses to tropical marine ecosystems, including shallow reefs and open water. Deeper and cooler water taxa survived. The evidence suggests global cooling as a cause. The Permian extinction saw massive losses to shallow marine and terrestrial ecosystems. There is some evidence for a series of extinctions in rapid succession. By the end of the Permian atmospheric warming driven by flood-basalt volcanism, global warming, marine anoxia and reduction of shallow marine habitats due to continental collisions caused global-scale geochemical disruption. Extensive areas of ocean are de-oxygenating nowadays due to nutrient enrichment from human activities, at the same time as we appear to be inducing climate change. The K/T boundary extinction is now credited to a massive meteor impact causing regional havoc and global change on top of another major flood-basalt event.

Extinctions, rebounds and radiations

Even the Permian mass extinction did not completely wipe out life. In the wake of such catastrophes ecosystems have recovered, rebounded and often repopulated by bursts of evolutionary activity characterised by the diversity of taxa and speed of change, called **radiations**. New species take up ecological roles previously occupied by the recently extinct. **Rebounds** and radiations vary in rate and outcome regionally and there can be a lag period after the extinction suggesting that ecological processes require some convalescence. The occasional major losses of **biodisparity** mean that radiations develop from an increasingly limited stock of fundamental lineages. Extinctions create opportunities for some survivors. The catastrophe that killed off the dinosaurs possibly opened up the opportunity for mammals that has resulted in humanity. The dinosaurs may even have an impact to thank, the large dinosaurs of the Jurassic appearing very soon after an apparent impact event at the end of the Triassic, marked by a characteristic iridium anomaly and fern spike.

The history of life is as much about extinction and loss as diversification and creation. Historic extinctions teach us important lessons. Many of the mass extinctions appear to be the result of multiple events, for example the Cretaceous extinction occurs when the ecosystem appears to have been already damaged by flood-basalt disrupting global climate and then an major impact creating a double-whammy. An unfortunate effect of human activity is that we create these multiple, simultaneous threats, for example climate change or ozone depletion degrading systems, and then specific damage such as local pollution of habitat destruction wiping out stressed populations.

Extinction may be perfectly natural and a creative force in the history of life but extinction and environmental change nowadays are widely seen as symptoms of a problem, not a natural process. What evidence is there for a biodiversity crisis in early twenty-first century Earth?

The current crisis

Miners' canaries

In 1973 a new species of frog was discovered living in the rock pools of montane forest streams in the Conondale and Blackall ranges of Queensland, Australia. Whilst trying to move a captive frog Queensland Museum staff were astonished to see the beast vomit up six wriggling tadpoles. Incredulity greeted reports that this new species, soon named the southern gastric-brooding frog (*Rheobatrachus silus*), nurtured its tadpoles in the female's stomach. The tadpoles survived in the potentially lethal stomach environment apparently due to switching off maternal digestive secretions with a specific prostaglandin released in mucus. In 1980 A\$23,000 was invested in pharmacological research investigating these inhibitory mechanisms with a view to human medical uses. Soon more was known about this frog's morphology than any other Australian amphibian. In the mean time the frog became a potent symbol for campaigns to protect the Conondale ranges, which are increasingly threatened by logging, habitat fragmentation and aquatic pollution. Local and international recognition of the danger to this remarkable amphibian saw its use on T-shirts and car stickers. A three-year study into the impact of logging was commissioned to begin in 1982, building on intensive surveys between 1976 and 1981. This frog was a classic example combining so many themes of biodiversity: a previously unknown species of the tropical forest, direct interest for biomedical benefit and a subject of local and international conservation campaigns (see Figure 1.13). This was a frog with a future.

Figure 1.13 Car bumper sticker using the Gastric Brooding Frog as a flagship species for conservation, in this case against the threat from logging of forests

Source: Redrawn from Taylor M. J. (ed) (1983) *The Gastric Brooding Frog.* Croom-Helm, London.

The last wild gastric-brooding frogs were seen in 1979. In 1980 and 1981 none were found. The gastric-brooding frog is now regarded as extinct, although you can find images on the new digital web-based Noah's Ark called ARKive (http://www.arkive.org).

The loss of this frog may be a natural, background event, unusual only in so far as we witnessed its passing. However, since the early 1970s, extinctions of amphibians have drawn attention to such losses. Examples have occurred globally and from different habitats. The golden toad (*Bufo periglenes*) of Costa Rica, famed for the brilliant orange colour of the males and intensively studied from the early 1970s, crashed to apparent extinction between 1988 and 1990. Common species also show population declines, e.g. western spotted frog (*Rana pretiosa*) is now extinct from a third of its range in the United States. The pattern of losses suggested that a significant event was underway. Losses were simultaneous and global; they included both range contractions and extinctions and extinction of species on well-managed nature reserves, the latter a particular worry suggesting that 'conservation as usual' on dedicated reserves was not enough to protect biodiversity. These losses of both the intensively studied and the common came to world-wide attention following an American National Research Council Workshop on declining amphibian populations, held in Irvine, California, in 1990. The workshop explored four themes:

- evidence of amphibian extinctions, population and range declines from around the planet
- evidence that such losses represented a genuine event that could be distinguished from natural (background) population fluctuations

- amphibians as indicators of general environmental degradation
- identification of any single cause.

There appeared to be many causes, sometimes interacting, and the problems of pinning these down were compounded by media coverage including suggestions of frog-napping by space aliens with a taste for amphibians. However, recent reviews suggest that several mechanisms are at work. First, there are comparatively well-understood ecological processes, impacts of introduced species, over-exploitation and changes to land use. A second set of factors are less well understood: global change (intensity of ultraviolet (UV) light, chemicals) and emerging infectious diseases, e.g. a specialist Chytrid fungi *Batrachachytrium dendrobatidis* and a virus *Ranavirus*. The fungus has been associated with nearly all frog declines studied in Costa Rica, including the loss of the golden Frog, and is known to have killed off the last few sharp-snouted frogs (*Taudactylus acutirostris*) when these were brought into captivity in a last-ditch attempt to save them. In 2001 the Global Amphibian Assessment Project was launched to pull together the disparate studies and attempt a planetary scale synthesis of the status of amphibians. In 2004 the bleak results were published. At least 43 per cent of all amphibian populations were in decline, with 9 extinctions and 113 species which 'can no longer be found'. Of the 5743 species, 427 are in the highest IUCN category of threat, 'critically endangered', and this is probably an underestimate since many species are too poorly known to reach a judgement. Nor could a single definitive cause be identified. The collapse of the Amphibia remains a startling but enigmatic event.

Amphibians may be an example of what have been termed **miners' canaries**, from the use in coal mines of caged canaries, which would keel over in the presence of poisonous gases, giving the miners time to escape. The amphibian losses are one of several miners' canaries, an advance warning of impending danger. Amphibians might be very good candidate canaries. A permeable, exposed skin and open eggs readily interact with the environment. Complex lifestyles risk disruption at many stages from different factors. Reliant on freshwater at least occasionally, their survival reflects wider catchment health. The consensus was that the recent declines were a remarkable event, not over and above the existing biodiversity crisis but a very clear signal of environmental degradation.

The accolade of miners' canary was widely popularised by Niles Eldredge's (1992) book. His exploration of the reality of recent extinctions opens with the observation that migratory songbird numbers have declined in the northern hemisphere. Eldredge suggests that these birds fulfil the same role on a global scale as the canaries of old. Songbird losses in North America echo the message of the frogs, except that one ultimate cause may be apparent. US Breeding Bird Census returns show local extinctions, population decline and range contractions of warblers, thrushes and flycatchers. These birds are forest specialists. The damage is done by fragmentation of woodland into patches below a critical size threshold. Small woodland remnants were exposed to predators and the egg parasite cowbird, which hunt along the woodland edges, scouring the whole patches more effectively. In the United Kingdom, the government's use of bird populations as one of fifteen headline measures of quality of life is an overt recognition of their role as indicators of the health of our environment. Declines include very familiar birds such as the house sparrow (*Passer domesticus*), whose numbers have fallen from 12 million to 6–7 million since the early 1970s, especially in towns and cities. The decline appears to result from decreased food resulting from changed agricultural practices. Sparrows tend to stay around the colonies where they were born, so declining or extinct populations are not topped up by new arrivals. This once common bird, taken for granted, has proved a miners' canary, its decline an early sign of the impact of changing land use.

Whole ecosystems may also respond as canaries, most famously the bleaching of coral reefs and collapse of temperate fisheries. Corals are members of the animal phylum Coelenterata, along with jellyfish and anemones. Coral reefs are built by corals that can secrete a calcium carbonate exoskeleton. Not all corals build massive reefs. Those that do are restricted to the tropics with average monthly sea temperatures no lower than 20°C all year round. The variety of corals is matched by the richness of other life dependent on the reef. Yet all this massive ecological engineering depends on a symbiosis of coral polyps with single-celled algae, called **zooxanthellae**, living inside the cells of the polyps' guts. Photosynthesis by the algae provides most of the corals' carbon supply and polyp waste is recycled. In return algae receive nitrogen and phosphorus nutrients scarce in the surface waters of the tropics, from food snared by the coral. The photosynthesis also creates an alkaline local environment allowing the calcium carbonate accretion (recent evidence suggests that the algae may perhaps be the dominant partner controlling calcification to enhance photosynthesis). Corals without symbiont algae can build skeletons but at terrible energetic costs and massive reefs are impossible.

Coral bleaching results from the loss of zooxanthellae. Algal cells can degenerate or be expelled. In extreme cases polyp tissues containing algae detach completely. The corals lose their colour and die leaving the bleached, white skeleton. Knock-on effects include loss of fish species due to habitat destruction and blooms of toxic algae, with economic losses rippling through to humans. Bleaching has been occasionally recorded since 1911 as a response to localised stress in hot, calm weather but from 1979 massive regional bleaching episodes have been recorded in every year (except 1981, 1985 and 1991). Bleaching events become more severe every year, affecting more sites, over greater depths and wider areas, resulting in greater mortality. The 1998 and then 2002 bleachings on the Australian Great Barrier Reef were the worst on record, breaking out along all 2000 km of the reef, affecting 42 per cent of sites in 1998 and 54 per cent in 2002. Many causes have been suggested: disease (no evidence), oxygen toxicity (but oxygen production tends to decline at bleaching episodes), UV light (but some bleach beyond UV penetration depth), pollution (but pristine sites have bleached) and heat plus intense light (some evidence for local impacts). One factor is consistently associated with mass bleaching, raised sea-water temperature. A mere 1°C increase over the normal long-term warm season average is sufficient to induce bleaching. The stress depends in part on the synergy between temperature and length of exposure, for example the Great Barrier Reef studies suggest that below 28°C there is no bleaching but twenty days at 30.3°C or ten days at 30.8°C provoke bleaching. Because of the link between thermal stress and bleaching the responses of coral reefs have been cited as early evidence of general environmental stress caused by global warming. Projections of sea temperature rises of 1.5–5°C over the next one hundred years have been suggested. Models of Great Barrier Reef bleaching suggest that an additional 1°C rise would increase the extent of bleaching events to over 80 per cent of the area. Rising sea levels and increased storms would be an additional burden to reefs whose resilience has already been degraded by over-fishing, pollution, sedimentation and disease. The bleaching episodes since 1979 may be an early warning, a whole ecosystem miners' canary, the first to register the effect of global warming (see Figure 1.14).

Losses of taxa and ecosystems in crisis are serious problems in their own right but the coincidence of events since the mid-1970s, their acceleration and the underlying message of planetary health in collapse as impacts cascade through natural systems setting off further disruptions suggests a deeper crisis. The miners' canaries have become too ominous and spectacular to ignore. The damage to natural ecosystems is no longer seen as merely symptomatic, evidence of damage already done, but as feeding back to precipitate yet more danger.

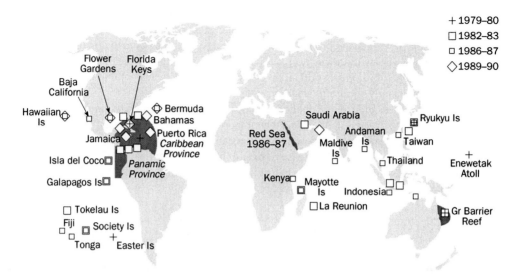

Figure 1.14 Global mass coral bleaching incidents from peak years, 1979–90

Source: Redrawn from Glynn (1991).

An end to evolution

We can all name an extinct animal. An awareness of imminent extinction is a familiar theme dogging the tail-end of most wildlife documentaries. The history of life on Earth is as much a tale of extinction as any other process. Extinction is so familiar, so natural that it begs the question, what are we worried about? Our concern may stem from a faddish environmentalism or an understandable but self-centred desire for planet Earth to remain much as it is, despite the haunting message of change from the ghosts of lost worlds. Extinction is not the problem.

The threat to biodiversity arises if extinctions due to human activity are qualitatively and quantitatively different from natural events. In addition the opportunity for natural rebound and recovery may no longer exist. Recent extinctions take place so rapidly, may be so widespread and cause such disruption to ecosystems that a global catastrophe snowballs. Human activity creates the double-whammy stresses associated with historic mass extinctions, for example the global stress to coral reefs apparently linked to climate change with local stresses such as over-fishing overwhelming the natural resilience of the ecosystem and causing catastrophic destruction. This combination of losses with the ferocity of a mass extinction with no chance of recovery has been termed the end of evolution (Ward 1995). Citing the Permian mass extinction as the first major event and the K/T catastrophe as the second, Ward believes that a third event, starting with glaciation some 2,000,000 years ago and accelerated by humanity is already underway. The combination of a climatic first strike (glaciation) causing broad disruption followed up by global damage wrought by humans to surviving ecosystems as analogous to a massive meteor impact is frightening.

Other scientists disagree and believe losses at the end of the twentieth century are not much different to natural rates. The idea has achieved mythical status. One critic has dismissed tales of a current mass extinction, as a case of scientists not letting the evidence get in the way of such an important idea.

The very word biodiversity can be abused. In 1996 Professor Arthur Obel, Kenyan Chief Scientist to President Daniel Moi, marketed a cure for HIV/AIDS, called Pearl Omega, at

£350 a bottle under a company name of Biodiversity. Launched with backing from the state press, Obel commented 'We have hit the jackpot!' whilst dismissing calls for evidence that it worked or of its nature and despite a ban on sales from the Health Ministry. Pearl Omega was withdrawn from sale later in the year following international press coverage. In the developed world biodiversity has become such a popular buzz word that articles increasingly sneer at this jargon.

The seriousness of recent extinctions can be judged only against past patterns. We need to understand what ecosystems do and their importance, if any, for general planetary health. We need better estimates of how much life shares the planet and the world's special biodiversity hot spots, accurate measures of current losses other than gaudy telegenic fauna and how much we can afford to lose before the crisis becomes terminal. We need to unravel the patterns and processes of biodiversity.

Summary

- The term biodiversity was coined in the 1980s and rapidly popularised, combining aspects of science, economics and politics.
- Life on earth is an ancient phenomenon. Earth is highly conducive to life. Life has survived massive global crises. Most species that have lived are now extinct but life has always recovered from previous crises.
- Recent species extinctions and ecosystem collapse suggest a new crisis, equal in scale to historic mass extinctions, may be taking place.

Discussion questions

1   Why did the concept of biodiversity catch the public imagination in the 1980s?
2   What messages, worrying or reassuring, does the history of life on Earth carry for biodiversity today?
3   Are there any miners' canaries in your country?
4   Did your country celebrate the International Day for Biodiversity on 22 May? Find out at http://www.biodiv.org/

Further reading

See also

Definitions of biodiversity, Chapter 3 pp82–83.
Current losses of biodiversity, Chapter 4 pp127–138.
International treaties, Chapter 5 pp174–279.

General further reading

Conway Morris, S. (1998) *The Crucible of Creation: The Burgess Shale and the Rise of Animals.* Oxford University Press, Oxford.

Eldredge, N. (1992) *The Miner's Canary.* Virgin Books, London.
Historic and recent extinctions and evidence for current crisis.

Gould, S.J. (1989) *Wonderful Life.* Hutchinson, London.
Gripping account of the Burgess Shale Fauna and wider lessons for our interpretation of the history of life on Earth.

Lovejoy, T.E. (1980) *Changes in Biological Diversity*. In G.O. Barney (ed.) *The Global 2000 Report to the President, Vol. 2 (The Technical Report)*. Penguin, Harmondsworth.
Founding paper describing biological diversity and including estimates of extinction rates as forest habitat declines.

Mooney, H.A., Lubchenko, J., Dirzo, R. and Salsa, O.E. (eds) (1995) *Section 4. Biodiversity and Ecosystem Functioning: Basic Principles* and *Section 5. Biodiversity and Ecosystem Functioning: Ecosystem analyses*. In V.H. Heywood (ed.) *Global Biodiversity Assessment*. Cambridge University Press for UNEP, Cambridge.
Very thorough, detailed and effective review of this major topic.

Norse, E.A. and McManus, R.E. (1980) *Ecology and Living Resources Biological Diversity*. In *Environmental Quality 1980: Eleventh Annual Report of the Council on Environmental Quality*. Council on Environmental Quality, Washington, DC.
Founding paper describing biological diversity; its wide horizons explicitly include philosophical and social topics.

Raup, D.M. (1993) *Extinction: Bad Genes or Bad Luck?* Oxford University Press, Oxford.
Review of historic extinctions, including mass extinctions, their causes and frequency.

Ward, P. (1995) *The End of Evolution*. Weidenfield and Nicolson, London.
Historic and recent extinctions and biodiversity.

Wilson, E.O. (ed.) (1988) *Biodiversity*. National Academic Press, Washington, DC.
The book of the 1986 National Forum of Biodiversity.

Other resources

Amphibian Declines (special issue). *Diversity and Distributions*, (2002), 9(2).
Whole edition of this journal devoted to amphibian declines and exploring up-to-date patterns and causes.

Convention on Biological Diversity: http://www.biodiv.org/
The Convention's own website and gateway to a wealth of information.

United Nations: http://www.un.org/
Well worth checking to keep up with the continuing saga of biodiversity in the context of global politics.

② The creation of biodiversity

Patterns of biodiversity are created by ecological and evolutionary processes. Lessons from ecology and evolution are important for conservation. This chapter covers:

- **Origins and remits of ecology and evolutionary science**
- **Ecological patterns and processes**
- **Evolution and the diversification of species**

Biodiversity is the outcome of evolutionary and ecological processes. True but glib. The university libraries where I work stock 1.8 km of shelves stuffed with books and journals dedicated to these topics. The purpose of this chapter is to provide a secure foundation from which you can explore these themes without feeling like you are swimming through blancmange.

The ecological domains are defined by the **population**, **community** or **ecosystem** and their interactions with the environment. The **phenotype** is central. Relevant ecological processes range from the autecology of individuals through to whole ecosystem function. The patterns and processes have an impact on genetic diversity but as a consequence. Ecological patterns and processes within communities and ecosystems may ignore the genetic domain which is manifested by species. The evolutionary domain is defined by a focus on genetic processes, patterns and consequences and the variation they create. The **genotype** is central. Relevant evolutionary processes range from adaptation (as genetic selection in response to the environment) through microevolutionary changes within a species to macroevolution, speciation and extinction. These changes may be driven by the environment but the genetic diversity remains the theme, linked to the species. Some genetic processes (e.g. mechanisms for mutation and variation) could function without any environmental influence.

Evolution, ecology and the origins of confusion

The origins of ecology and evolutionary science

A fascination with the origins, patterns and causes of nature's variety can be traced back at least as far as Greek texts, most famously Aristotle's (fourth century BC) *Historia Animalium* listing over 500 species followed by Theophratus (fourth to third century BC) with a *Historia Plantarium* describing plants and their uses. From the start attitudes to nature split between a vision of purposeful, harmonious systems versus nature as nothing more than individualistic bits linked like a machine. These classical writers may be an artificial horizon, a sort of Cambrian explosion of surviving texts. Ancient cave paintings

Plate 7 Biodiversity promotes healthy ecosystems. Using sophisticated indoor growth chambers, called Ecotrons, ecologists at the Centre for Population Biology have shown that a diverse mix of plant species recycles gases and water more efficiently than a similar biomass of few species. Photograph COI/CPB

(e.g. Vallon-Pont-d'Arc), unintelligible but haunting murals of humans and animals, hint at an awareness of the interactions of nature in humanity's archaic history. Medieval and pre-Enlightenment scholars maintained a robust interest in nature's variety, interactions and importance. The thirteenth century *Summa Theologie* of St Thomas Aquinas analyses the orderliness of nature, including the sentiment that could be drawn from the most recent biodiversity texts on ecosystem function: 'It is better to have a multiplicity of species than a multiplicity of one species' (see Plate 7). Other insights were less useful: biodiversity included many dubious mythical wonders.

The European Enlightenment revolutionised attempts to classify the natural world. New classifications used the physical resemblance of animals and plants to catalogue and to create an order, in keeping with the growing intellectual confidence of the age. Older magical, symbolic, sometimes medical categorisations were lost. Many Enlightenment systems were firmly grounded on a vision of divine creation with species as immutable types, any new species at best filling up preordained pigeon holes. Most famous is the work of Carolus Linnaeus, a Swedish biologist, who devised the system for applying classifying and naming species still in use nowadays. The themes of other natural historians are familiar in present-day biodiversity research. In the 1700s John Harris published *Lexicon Technicum*, describing natural history as 'a description of any of the natural products of the earth, water or air, such as beasts, birds, fishes, metals, minerals, fossils, together with such phenomena as at any time appear in the material world, such as meteorites'. In 1749, Georges, Comte de Buffon, Keeper of the Royal Zoological Garden in Paris, started publication of his *Histoire Naturelle*, intended as an inventory of the Earth and its wildlife, highlighting patterns of geographical distribution, adaptation and migration, a global biodiversity assessment of its day. The role of species in maintaining a healthy environment was documented: for example, the Reverend Gilbert White wrote on earthworms in letters later published as *The Natural History of Selbourne* in 1789 and Richard Bradley, an eighteenth-century horticulturist, commented on the value of birds for pest control. The interdependence of species was understood by John Ray, for example, who wrote *The Wisdom of God Manifested in the Works of the Creator* in 1691.

The accumulation of knowledge continued through the early nineteenth century. Finally came the publication in 1858 of Charles Darwin's and Alfred Wallace's theories of evolution by natural selection, informed by the distributions of living species, fossil records and a mechanism for evolution driven by day-to-day contest for limited resources. The only major component missing was an understanding of genetics. Even if not yet named evolutionary and ecological science appeared to have had a very sound gestation, but the reverse was true.

Modern ecology: illegitimate origins and confused childhood

The disparate heritage of ecology caused problems defining the legitimate territory of this science. The term ecology (then spelt oecology) was coined by the German zoologist Ernst Haeckel in 1866, in part to categorise the topic as a mere aspect of physiology and distinguish it from biology. Ecology was an awkward combination of the traditions of natural history (everything from grotesque menageries through to the ideas of Darwin) and the other major thread of nineteenth-century natural biology, physiology. At various times both traditions subsumed or abhorred the newcomer. In 1893 Sir John Scott Burdon-Sanderson, President of the British Association for the Advancement of Science, distinguished ecology as one of the three divisions of biology (physiology and morphology were the other two), with ecology defined as 'a philosophy of nature'.

Ecology is not a radiation of ideas from a secure central core, nor a synthesis of ideas into a greater whole. It is an umbrella of concepts. Three blights (no secure pedigree, multiple parents and a subject matter unloved by the great biological themes of the late nineteenth century) got the subject off to shaky start.

A fourth source of confusion, in part a reaction against this sense of illegitimacy, soon developed. During the early twentieth century there was an emphasis on making field studies rigorously scientific to bolster ecology's status, to understand nature all the better to master and improve upon it. There was a desire to make ecology coherent by developing a quantitative, mechanistic, reductionist approach, as if nature could be stripped down like a big machine, echoing the rigours of physics and chemistry. Ecology had transferred its interest from explanations dominated by history (the legacy of **biogeography**, dispersal and evolution) to explanations that forgot time. Ecology as a science had no clear past and now ignored historical time as a process. G.E. Hutchinson's famous paper of 1959, 'Homage to Santa Rosalia or why are there so many kinds of animals', is a revealing example of the shift in emphasis. Specifically addressing global animal species' richness, Hutchinson lists factors to consider for a theory to predict the total: food chains, plant–animal interactions, complexity and stability of ecosystems, productivity, size, environmental rigour, competition and the environmental mosaic. The paper omits historical ecology.

Ecology was riven by rival schools of thought – population/community ecology concentrating on species and populations versus ecosystem ecology analysing the cycles and flow of energy and resources with a sense of regulation and balance. Alternative visions included communities as integrated super-organisms versus individualistic assemblages. This contest between ecological systems as benignly self-regulating or driven by selfish contest echoed the Enlightenment, even the Greek, debates between a vision of nature driven nature by purpose versus trial and error. Then, just when ecology was trying to look grown up by talking in numbers, growing public awareness of environmental problems in the 1960s saw the term ecology purloined as a general social and political creed. Many scientists were wary of or annoyed by the blurring of their science with environmental politics. The emergence of a science of biodiversity has to some extent healed this rift.

The ecology of biodiversity

The search for patterns and processes

Robert MacArthur studied under Hutchinson, driven by the same question as his mentor: why are there so many species of animals? MacArthur's work became a dominant influence in late twentieth century ecology, immortalised in the MacArthur–Wilson Theory of Island Biogeography (the very same Wilson who edited *BioDiversity* in 1988). In 1972 MacArthur's book *Geographical Ecology* was published, written when terminally ill. The introduction opens: 'To do science is to search for repeated patterns'.

Later, Chapter 7, entitled 'Patterns of species diversity', begins with the frightening litany of patterns, tropical to temperate, land to sea, plant and animal and frets:

> will the explanation of these facts degenerate into a tedious set of case histories or is there some common pattern running through them all? A very brief review of explanations that naturalist have proposed will help put the discussion in perspective.

This section faces the very same challenge to produce a coherent review of the many ecological creating diversity and controlling its distribution: first, a hierarchy of factors differentiating three broad types of ecological factors, primary influences of time and space, second, templates such as physical habitat structure or patterns common to food webs, and third, intimate processes depending on individual species actions and attributes.

Primary factors, geographical and physical

Primary ecological factors are large-scale global or regional physical influences. They determine the general environment within which the other ecological factors create local, intimate detail or disruption:

- *History and age*: biodiversity is generally greatest in oldest ecosystems.
- *Gradients*: biodiversity changes across environmental gradients (e.g. latitude, altitude, depth, aridity and salinity).
- *Area*: biodiversity increases with increasing area.
- *Isolation or islandness*: biodiversity decreases with increasing isolation.

Ecological templates

Biodiversity at regional or smaller scale suggests patterns or structures to which species diversity adheres or is driven to fill up. Structures may be literal (e.g. the physical architecture of a habitat) or apparent (e.g. rules governing how species link together in food webs).

- *Productivity; available energy*: globally biodiversity increases with productivity but locally increased or decreased productivity typically lowers biodiversity.
- *Scaleless patterns*: ecological factors such as area or productivity have a measurable scale (e.g. diversity per hectare or relative to solar energy). Scale-less patterns include the consistent patterns found in food webs, patterns of species richness relative to population abundance and body size.
- *Habitat heterogeneity and patchiness*: the greater the variety of types of habitat, the greater the diversity of species.
- *Habitat architecture*: diversity increases with increasing architectural complexity of the physical habitat.

Interactive, intimate factors

Primary factors and ecological templates typically promote or reduce biodiversity in a predictable way: biodiversity increases with increasing area, or decreases with decreasing habitat complexity. The third set of factors can enhance or suppress diversity depending on the spatial and time scales. They are characterised by the local, intimate scale at which they operate, directly dependent on the attributes, behaviours, interactions and susceptibilities of the species involved.

- *Succession*: successional processes comprise the arrival, establishment and replacement of species through time. Changes may be driven by environmental agents and species altering habitats.
- *Interactions between species*: interactions between species influence short-term, local diversity and power evolutionary diversification as winners and losers, exploited and exploiters struggle for dominance.
- *Disturbance*: local physical destruction, as against regional, climatic variations linked to gradients, may destroy diversity. Equally, disturbance may be frequent enough to prevent a limited gang of species monopolising a habitat but not so frequent that everything is wiped out.
- *Dispersal and colonisation*: dispersal (active or passive) and colonisation (arrival plus survival) depend on the attributes of individual species and represent the species interface with the primary factor of isolation.

Primary factors

History and age

The biodiversity of an ecosystem will depend on the time it has existed and previous history. History as an explanation for present-day patterns of biodiversity was the factor that twentieth century ecology forgot. In part this was because history seemed little more than a *Just So* story; species are where they are because that is where they were, the patterns untestable without the use of a time machine. Modern biodiversity is inevitably contingent on past events. Increased understanding of the Earth's history has sharpened our awareness of history as a factor underlying existing biodiversity.

The distribution of related taxa across widely separated continents, especially in the southern hemisphere, had proven a challenge to nineteenth century biologists. The scatter was explained by lineages jumping across oceans. The revelation of continental drift during the 1960s permitted an alternative view. Present-day distributions are the result of two historic processes: the **origination** of lineages on a united super-continent and the subsequent split and **continental drift** of fragments. This physical splitting is **vicariation** and the resulting diaspora called **vicariance**. Vicariance patterns depend on where the common origin range was, how the now separate fragments once joined and where splits arose. Large-scale **dispersal of lineages**, which might include some jump dispersal, is responsible for range expansions, which then might suffer vicariance. In addition to continental drift a gradual expansion of the Earth resulting in a stretched crust may be a historical factor forming vicariant patterns (see Figure 2.1).

A second major historical influence are **refugia**, discrete areas acting as sanctuaries during times of global stress. They form the shrunken fragments of larger **biomes** and expand, reforming into wider swathes during benign periods. The idea was developed from the fate of tropical rainforests during the Pleistocene glaciations of the northern hemisphere, though could apply to many global biomes. Not only do the refugia conserve existing biodiversity, but also the fragmentation may increase net diversity due to speciation, the new species surviving later reunion of refugias. The great age of tropical rainforests has been used as an explanation that seems to contradict the refugia hypothesis, the **stability–time hypothesis**. Tropical rainforests are ancient, stable habitats, allowing prolonged accumulation of species. Rainforests have existed for tens of millions of years. Pleistocene glaciations are much more recent. Both historic processes have added to their riot of biodiversity.

Age and history are important: massive physical processes such as continental drift and glaciation linked to dispersal of lineages and evolution explain the broad patterns of biodiversity distribution and endemism. Large-scale biodiversity differences between continents are contingent on the past. We would not have the unusual faunas of Australia and Madagascar if it were not for their history.

Gradients

Biodiversity does not litter the planet at random but typically changes across environmental gradients (see Figure 2.2). The most famous gradient is the increase in biodiversity from high latitudes, the poles, to low latitudes, the tropics. Trends of light, temperature, climate and seasonality facilitate diversity. Increased diversity towards the tropics occurs in land mammals, birds, reptiles, amphibians, insects, snails, clams, trees and plankton. Marine fish diversity increases towards the tropics but some marine trends are sharply stepped, linked to distinct fronts between massive water fronts. Exceptions to the trend include some marine groups, e.g. penguins, perhaps the result of their south polar origins, although the diversity of ecologically similar Auks in high northern hemisphere waters hints at other influences. Terrestrial exceptions include conifers, which are most diverse in temperate latitudes. The decrease of diversity with altitude is a response to similar environmental conditions, with similar exceptions of conifers most diverse at mid-altitudes of mountain ranges.

Biodiversity changes with depth underwater but the deep sea remains largely unexplored, so generalisations are risky. Diversity seems to decrease with depth, certainly in high diversity habitats such as coral reefs. Other marine systems show greatest species richness at intermediate depths. Samples of deep-sea floor life increasingly hint at a rich variety. Depth is not simply an underwater analogue of altitude but a response to the synergy of light, temperature, pressure and dissolved substances such as oxygen. Geographical gradients also include aridity and salinity. Diversity is lower in arid regions compared to wet zones of similar area, altitude and distance to the sea. Aquatic diversity

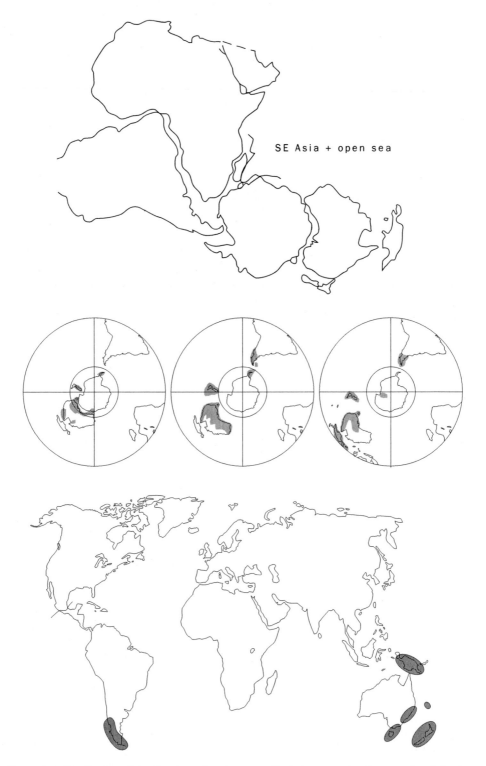

SE Asia + open sea

Figure 2.1 Distribution of the Southern Beech, genus *Nothofagus*, a classic example of vicariance. (a) The modern continents when joined as Gondwanaland; (b) The subsequent separation of southern continents. The shaded areas show areas with fossil evidence for the presence of *Nothofagus*; (c) Modern distribution of *Nothofagus*, apparently very disjunct but in regions historically close together

Source: Redrawn from Tarling (1972). Another Gondwanaland, *Nature*, 238, 92–93 and Hill (1992) Nothofagus: evolution from a southern perspective, *Trends in Ecology and Evolution*, 7, 190–194.

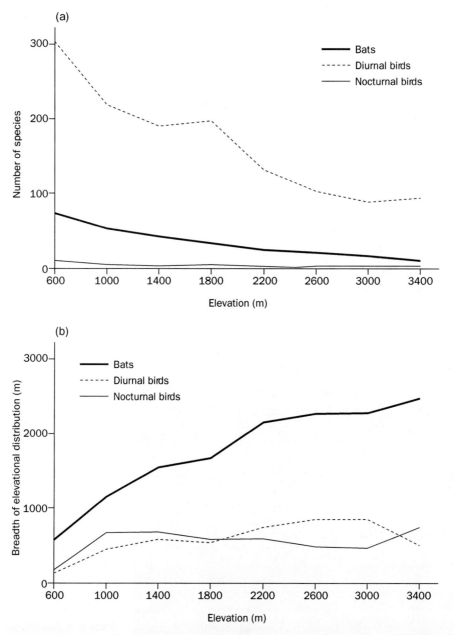

Figure 2.2 Numbers of bat and bird species recorded over an altitude gradient in the Peruvian Andes. Numbers of species decline with increasing elevation. As altitude increases bat species show wider ranges, perhaps having to roam further to survive, in turn restricting numbers that may live together. Oddly birds do not show this trend, so different taxa may be affected by gradient in different ways

Source: Redrawn from Graham (1990) Bats versus birds: comparison among Peruvian Volant vertebrate faunas along an environmental gradient. *Journal of Biogeography*, 17, 657–668.

is highest in either sea or freshwater but decreases sharply at brackish concentrations in between (see Plate 8).

Gradients can be divided between regulation gradients where the controlling factor is not used directly (e.g. altitude) or resource gradients where it is (e.g. nutrients), altering availability. Although biodiversity gradients are so striking, explaining how they arise is difficult.

Most attention has focused on the polar–tropical gradient. **Time** may be a factor. Pleistocene glaciations smothered polar and temperate regions causing extinctions. The recent retreat of the ice has not allowed time for species to recolonise higher latitudes. Meanwhile the tropics survived, allowing time for speciation. So time as a factor is linked with disturbance. This combination is encapsulated by the idea of ecosystem **stability**. The stability of ecosystems (combining **resistance**, the ability to absorb impacts without suffering change, and **resilience**, the ability to recover from changes that are inflicted) was seen as a dominant influence. Stability encouraged complexity which in turn reinforced stability. These ideas are encapsulated as the **stability–time hypothesis**. Contrarily disruption and instability has been cited as promoters of tropical diversity.

Plate 8 A diversity gradient with altitude. Creag Fhiaclach in Scotland is believed to have the only natural tree line in Britain. The Caledonian pine woods of the glens give way to birch then montane herb and grass communities as altitude increases and conditions become increasingly harsh.

Contraction and fragmentation of tropical forests followed by expansion and fusion in response to glacial cycles may have provided opportunity for local speciation in the fragments, the diversification surviving once the forests expanded and coalesced.

Habitat heterogeneity is another ambiguous factor. Tropical forests and coral reefs are more heterogeneous, providing more refuges and opportunities. But rainforests and coral structures are a response to gradients, not a cause, although they in turn provide an arena for the diversification of other life dependent on them.

Tropical diversity seems closely tied to benign conditions neither too harsh or unpredictable. This link has been termed the **favourableness** of the environment. Favourable is tricky to define, not least because a harsh environment is not harsh for taxa adapted to live there. Tropical forests are warm but not too hot, wet but not waterlogged and not highly seasonal. These generous conditions permit high productivity, high biomass and high diversity. Low diversity occurs in either constant but extreme (e.g. deserts) or fluctuating (e.g. estuarine) habitats.

Favourableness promotes **productivity**. The low diversity of large, old, predictable (all characteristics supposed to promote diversity) habitats such as deserts suggests restrictions on speciation or colonisation, perhaps due to low productivity. Fundamental constraints such as temperature and light act as severe limitations. High tropical diversity requires that productivity is divided between more species not monopolised by larger populations of the same, few species found elsewhere. A diversity of specialist species may exploit the full range of productivity very efficiently, excluding a more limited variety of generalists. The favourable tropics cover large areas, ensuring that total productivity is high, effectively a bigger energy cake to be divided, sufficient to maintain the small populations typical of many rarer species, without undue risk of accidental extinction associated with tiny populations or ranges. This possibility is termed the **species–energy hypothesis**.

The lack of physical constraints in the benign tropics may permit more interactions between species, more intense, more specialised, more likely to reach a conclusion as arms races spiral and competitors outwit one another. The **interaction hypothesis** is awkward to demonstrate and circular: diversity begets diversity. Nonetheless, interactions once accelerated in favourable conditions may snowball.

An alternative to this tangle of favourableness, productivity and interaction is that tropical high diversity is nothing more than a variation on the theme of another primary factor, area. The tropics harbour greater diversity because they cover a greater area. Not only is their extent absolutely larger than other global biomes, but also northern and southern hemisphere tropics join, unlike the separation of temperate or Arctic zones.

Area

Biodiversity increases with increasing area. This rule of thumb remains an awkward ecological pattern to explain. A large area of polar habitat will contain many fewer species than a smaller area of rainforest; glib comparisons can be dangerous. The pattern may be little more than a sampling artefact since bigger areas act much like bigger nets, collecting up more species without any especially ecological mechanisms at work.

The link between area and biodiversity is enshrined in **species–area relationships**. Numbers of species increase with increasing area. Data depicting this pattern are commonly plotted to give a species–area curve, the increase of species with area typically slowing as the pool of species still to find is used. Species number and area are often plotted transformed into their logarithms, resulting in a plot approximating to a straight line rather than a curve. Straight lines are much easier to model as equations and many species – area relationships can be described by the equation $S = cA^z$. S is the number of species and

A the area; c and z describe how species numbers change with area; c is the number of species per unit area and then z adjusts the effects of area which may vary between different habitat types. The hunt for consistent patterns has focused on z. Very similar ranges of z, between 0.25 and 0.3, for different habitats have been cited as evidence supporting some fundamental ecological rule governing species diversity in relation to area but z can be much more varied (see Figure 2.3).

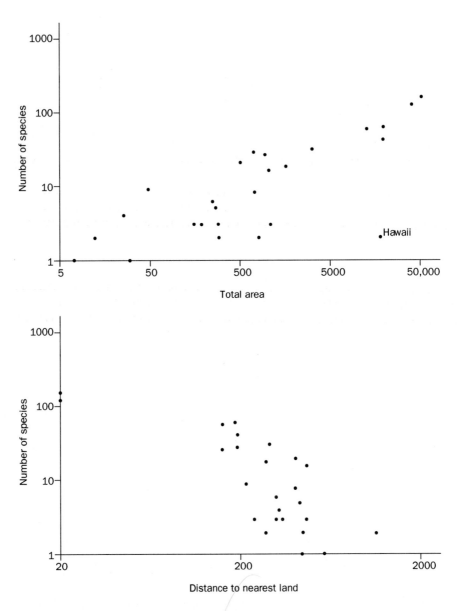

Figure 2.3 Species:area and species:isolation patterns for butterflies of the Pacific islands. Numbers of species found in different archipelagos are plotted against archipelago area and distance from nearest mainland. Species numbers increase with area, decrease with isolation. Note the lack of species in the Hawaiian islands, despite their total size due to extreme isolation. Area and isolation alone can explain 75 per cent of the butterfly species diversity of Pacific islands

Source: Redrawn from Adler and Dudley (1994) Butterfly biogeography and edemism on tropical Pacific Islands, *Biological Journal of the Linnean Society*, 51, 151–162.

Explanations for the effect of area on diversity vary from suggestions that the patterns are little more than **statistical artefacts** through to area as the predominant motor of ecological diversity. If all species were distributed at random, a bigger area would contain more, but since they do not occur at random, except perhaps at a local scale, this simplest explanation does not hold. Other non-random distribution patterns may generate species–area curves. Many ecosystems support total numbers of species and populations of each that follow a very particular pattern. Many species occur at low or moderate densities, a few at very low and a very few at very high abundances. Plot the numbers of species with populations in each category and the distribution of species across population categories fits a pattern called the log-normal. Different sized areas drawing their species from a total pool that matches a log-normal pattern can create a classic species–area curve. The important ecological processes are those that produce the log-normal pattern in the first place, not the area.

Larger areas contain a greater variety of habitats – **habitat heterogeneity**. Not only will larger areas contain more types but also the types will include increasingly different habitats. Combine the increased variety of habitats as area increases with the variations in detail at different scales and area's effect in determining species number may be the result of habitat heterogeneity.

The extent of an area will affect evolutionary processes, the opportunity for speciation and risk of extinction. Biodiversity differences between provinces and continents of varying size echo their history. The **species–time hypothesis** recognises the role of time, larger areas harbouring more species due to more opportunity to speciate and less risk of extinction.

Area remains a contentious factor. In an extremely thorough exploration of the ecology of diversity, Rosenzweig (1995) entitles a summary section 'A large piece of the puzzle: area effects', combining statistical, habitat heterogeneity and evolutionary processes. Other authors demean area as little more than a handy surrogate for the underlying mechanisms.

Isolation

More isolated habitats harbour a reduced biodiversity. Isolation is not simply the distance between a source of colonists and an island but the interaction of distance, species dispersal mechanisms and any special hazards disrupting access. Rosenzweig (1995) distinguishes mainland (essentially area) from island (isolation) patterns. Islands include not only literal oceanic isles but also separate patches of the same habitat scattered through the landscape. Genuine islands may be self-contained with species originating entirely from immigration and none depend on continual topping up by immigrants for survival. Mainlands may be populated by taxa that originated from speciation within the area. By this definition Hawaii is a mainland.

Island biogeography, epitomised by MacArthur and Wilson's Theory of Island Biogeography, depends on two processes: immigration, which varies with isolation from colonisation source to island, and extinction, which depends on island area. As more species colonise, the pool of potential colonists is used up. In the mean time extinctions increase, perhaps simply because every species has some chance of accidental extinction or maybe because of interactions between the increasingly cramped arrivals. An equilibrium diversity is reached once immigration and extinction cancel each other out. This may be a **dynamic equilibrium**, as species colonise, go extinct and recolonise. A **turnover** of species is common, though hard to measure since a species may be found, then go extinct and re-establish before the next survey, so-called cryptoturnover.

Island biogeography predicts that should an island be perturbed so that it harbours more or less species than the equilibrium and assuming the perturbation does not alter the island habitat, then the equilibrium will re-establish. Experiments following the fate of islands intentionally stripped of fauna suggest that not only does the equilibrium restore but also other features such as the general patterns of predator and prey species, the trophic structure, are rebuilt to mirror the pre-disruption status quo. However the actual species involved may differ. Such results suggest there are places to be filled, limits and opportunities for diversity.

Isolation and area create **source** and **sink** habitats. Sinks are patches in which a species is unable to maintain a viable population without topping up by immigrants. Sources are areas where a species' reproduction is sufficient for survival. The arrival immigrants to a sink patch is called the **rescue effect**.

Isolation and inter-patch processes have been developed as **metapopulation ecology**. A metapopulation is a set of populations in separate patches but linked by immigration (see Figure 2.4).

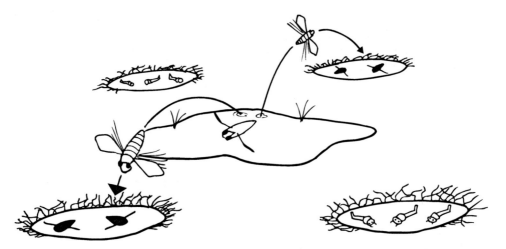

Figure 2.4 A metapopulation of greater waterboatmen, *Notonecta* sp. The boatmen survive in the permanent central (source) pond. Peripheral ponds may be colonised by flying adults, which then breed. Ponds not colonised contain prey species that are wiped out in the presence of the boatmen. The peripheral ponds dry out each year so that the ecological stage is reset. Note how habitat patchiness and disturbance, dispersal and predation all affect local diversity

Ecological templates

Productivity

On a global scale biodiversity increases with rising productivity, one factor influencing planetary biodiversity gradients. At a local scale productivity may limit or promote species diversity. At low productivities diversity is low. Some species simply cannot survive on the limited resources. This may be a particular limitation to specialists. At very high local productivity, diversity declines as a few, competitively dominant species, freed from resource constraints, monopolise the habitat. Other species are ousted, unable to resist the competition or perhaps simply unable to utilise the riches. The increased nutrients may also promote dynamic instability in **food webs**, the resulting species interactions leading to extinctions as populations fluctuate dangerously. At intermediate productivity levels there are sufficient resources to allow many species to survive but none may take over; this is the **intermediate productivity hypothesis**. Intermediate productivity may maintain diversity as the variety of life generates local habitat heterogeneity, both of physical

Plate 9 A species-rich marsh, Aberlady Bay, Scotland, a Local Nature Reserve. The marsh nutrient supply is productive enough to support a variety of species but not over-enriched (the fate of many wetlands vulnerable to agricultural fertiliser run-off) which reduces diversity as a few species monopolise the habitat.

structures (e.g. plant growth forms) and of resources such as nutrients or moisture. So productivity links to the resource needs of individual species and **competitive interactions**. Productivity has also been linked to **disturbance** in the **dynamic equilibrium model of species diversity**. In this model diversity is determined by the synergy of production (and its effects on population growth and competition) and **disturbance** (varying in frequency and intensity) (see Plate 9).

Productivity may also tie in with **scaleless patterns** of species diversity such as **species–abundance patterns**. If species diversity fits species–abundance patterns which are not purely ecological but much more general rules governing collections of things, and the productivity pie is sliced up accordingly, then the more pie (the greater the productivity), the more species will get a share, so occur in a habitat.

Local declines in diversity at increased productivity have been shown in terrestrial and aquatic habitats, especially where nutrient enrichment occurs. Enriched aquatic systems lose diversity as algae, able to respond quickly to increased nutrients, bloom and smother larger plants. Enriched grassland swards also lose species as a few coarse grass species take over. In both cases removal of nutrients can reverse the trend and local diversity recovers.

Scaleless patterns

Scaleless patterns include apparent correlations between body size, population abundance and number of species, plus rules for the structure of food webs. The relationships between species numbers, population abundance and body size have been a focus of recent work. Plot the numbers of species in a community against their populations and some consistent **species–abundance** patterns emerge. Models to explain these patterns focus on how species divide up available resources (resource apportionment) or statistical patterns that apply to collections of things in general be it ecological or otherwise. For example the 'geometric series' assumes a first species to establish monopolises a share of the resources, the second species in takes the same percentage of the remaining resources, the third the same percentage of what is left and so on until all resources are used up. Other models have no specifically ecological roots and make no assumptions about what the species do. Such models can just as well describe patterns of diversity in collections of any objects, whether insects, kitchen cutlery or chocolate bars. For example the log-normal distribution pattern arises when the abundance of items (in this case, populations of species) is the result of many random, independent influences.

Species–size
Species–size patterns take a similar approach, plotting the number of species against categories of body size, looking for consistent patterns and trying to explain these with general models.

Species–abundance–size
Species–abundance–size patterns have proved frustrating. There are some potentially important lessons for biodiversity and conservation, in particular if any consistent patterns linking abundance to size or diversity suggest that there are some population sizes simply too small to survive. It is possible to find ecological communities to fit each model but none are generally applicable.

Food webs
Consistent patterns in food webs (species linked as those higher up the web exploit those below for food) suggest that rules govern web structure. A species' position in a web is defined by how it captures energy. Primary producers trap the raw energy of sunlight, sometimes other sources, and convert this into stored chemical energy. Primary consumers including herbivores eat the primary producers and secondary consumers, predators and parasites, prey on the primary consumers. Since predators have their own enemies, there may be several levels of secondary consumers. Each of these categories is a **trophic level**.

Food webs show consistent patterns. First most contain no more than three or four levels. Longer webs can be found, more out of a love of the unusual. For example, some marine webs may go up to eight, assuming tortuous series of links from tiniest plankton to largest sharks, by way of four or five levels of secondary consumers eating one another. The consistent rule stays sound, most food webs reach three or four trophic levels. The second pattern is that omnivores, defined as a species that takes at least 5 per cent of its food from another trophic level than its main supply, are rare. Third ratios of numbers of predator to victim species, links per species, and links between different levels are also strikingly consistent in real webs. As diversity of one increases so does diversity of the others.

Limits to food web length were originally ascribed to **energy transfer** from one level to the next. Only 10–20 per cent of energy eaten is assimilated, the bulk being lost via excretion or metabolism, so available energy rapidly peters out up the chain, leaving too little to support any more levels. If available energy was the limiting factor, webs from more productive habitats should be longer but they are not. Only in the most unproductive habitats (e.g. terrestrial Arctic) is there evidence for truncation due to low energy. An alternative to energy limits is **dynamic stability** of webs as they become ever longer. Model food webs can be built to mimic realistic producer, predator and prey interactions, the fate of populations tracked over many generations. The longer the model food webs, the more unstable they become. As populations rise and fall, interactions ripple through the web and feedback lurches through the system. Long webs lose levels, collapsing back to more stable two, three or four levels. Dynamic models generate realistic web length limits, realistic predator–victim ratios, numbers of links and rare omnivory.

Energy flow and dynamics may be linked. In low productivity systems stability increases with an increase in productivity. In high productivity systems any further increase in productivity promotes instability. Taken together productivity may limit web length in low productivity habitats, instability in high productivity systems.

Realistic links and ratios can be created by a third, strikingly simple, theory, the **Cascade model**. Ignoring population dynamics and energy the Cascade model suggests that species are arranged in a hierarchy, perhaps by body size, so that each feeds only on those below and is fed on by those above. The total number of links per species is fixed, but how species link is randomised. The Cascade model predicts web length limits poorly.

Habitat heterogeneity

A varied physical environment harbours a greater biodiversity. Heterogeneity refers to the kinds, variety and distribution of different types of habitat. All environmental factors vary across time and space. The more variation, of size, age, type and amount of patch, the greater the diversity of life. Patches can be physically discrete areas or variations in the distribution of resources. This patchiness, coupled with changes over time, mitigates against a few species monopolising the environment. No species can be good at exploiting every environment. There is a trade-off, strong evolutionary pressures to optimise strengths whilst losing less successful attributes. The process will be self-reinforcing, with less and less to be gained from trying to exploit patches that other species can utilise more effectively. Even an apparently homogenous environment will become varied as different species colonise, altering their local environment. The result is that different patches will vary in their usefulness to species (see Figure 2.5).

Figure 2.5 Habitat heterogeneity. The greater the variety of habitats, in this case plant species, the greater the variety of species, in this case insects associated with each plant, although the overall quantity of habitat has not changed

Patchiness mitigates against monopoly by just a few species. Changes in one patch (losing out to a competitor) may be compensated for by a different fate in another patch (the victorious competitor wiped out by its predator). Local variation between patches results in regional coexistence of species. This is especially so if the ecology of separate patches is out of synch. For example, if disease strikes at a particular stage in a life history, populations will survive in some patches that are not at that vulnerable stage. Habitat heterogeneity has been shown to promote coexistence of predators and prey, e.g. many insect predators such as ladybirds can wipe out an individual patch of prey but cannot find all prey colonies hidden in different patches, and of competitors, e.g. many plant species which differ so little in requirements that competition would be severe coexist regionally scattered across separate patches. So the fate of individual patches links with **isolation** and **area** to create regional diversity. Temporal variation will also maintain diversity. Seasonal changes can upset the otherwise inevitable outcomes of interactions between species, resetting the ecological stage. Different species may suffer resource bottlenecks at different times of year, even if they are using the same resource (see Plate 10).

Plate 10 Habitat heterogeneity. This small stretch of river supports discrete patches of vegetation, submerged and emergent, as well as open water. The different patches each provide different habitat and alter the physical environment, increasing the local diversity of animals.

Habitat diversity (heterogeneity and complexity) and species diversity co-evolve. The richness of habitats in a region depends in part on the species to recognise their existence. Species discriminate the finer grain of habitats (be it patches or complexity) because selection forces them to. Natural selection finds the habitat diversity. Some regions with apparently very limited environmental diversity, e.g. the Fynbos of South Africa or Rift Valley Lakes, can support a surprising diversity of species (plants and fish respectively) that show a very fine-grained division of the environment.

Habitat complexity

The more complex the architecture of a habitat the greater the diversity of species. Complexity refers to the structural architecture of each habitat. The amount and shape of structural complexity in the environment will create opportunities for species. A structurally complex ecosystem (e.g. the three-dimensional architecture of a forest) provides more physical habitats for species than a simple one (e.g. an equivalent area of tidal mud). Habitat complexity sounds intuitively obvious but is difficult to untangle from other factors. For example in the forest–mudflat comparison above, the diversity of plants in an area of wood provides many more sources of food for herbivores (and in turn their enemies) than a mudflat (see Figure 2.6).

An important insight into complexity is that there is more of it at smaller scales. Many natural phenomena, whether soil structures or vegetation, are fractal. They do not have a finite size but measurements of dimensions such as area or edge increase the finer the scale at which they are measured. To a 1 m tall grazing deer a small herb like a buttercup is scarcely a mouthful; to a 5 mm long insect a buttercup is a complex climbing frame, with leaves, shoots, flowers and buds, each offering different resources. So the availability of microhabitats increases at smaller and smaller scales. As a result species biodiversity increases at smaller scales. There is more room for organisms of smaller body sizes. The biodiversity of terrestrial insects may in part be the result of this pattern, linked to the complexity and scale of the physical habitat that plants create. Many habitats are exactly the right scale for the diversification of insect-sized life (see Plate 11).

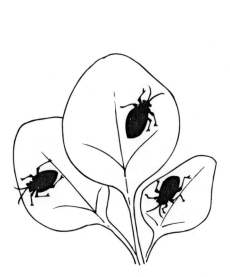

Figure 2.6 The greater the habitat complexity the greater the diversity of species. In this case the physically complex plant hosts more species. The variety of different sized and shaped spaces on the complex plant creates opportunities for more species, which vary in body size

Plate 11 Habitat complexity. The simple, straight leaves of the bur reed, *Sparganium erectum*, provide a less complex habitat than the branches of the water mint, *Mentha aquatica*. More species can exploit the greater complexity offered by the mint habitat.

Intimate processes

Succession

Diversity changes as species establish, interact and alter the environment. The biodiversity of a site increases over ecological time, immigration, establishment and loss of species creating a successional pattern. Succession is the non-seasonal turnover of species over time. The turnover is predictable with recognisable early colonists and late survivors. The progress of succession is directional, the sequence of early and late successional communities repeated across separate patches of the same habitat type in the locale. The directional, replicated changes may converge on a final community. Successional change comprises a sense of direction, of species turnover and of physical change as the biodiversity creates its own structures and alters the environment.

Early twentieth century ecology was mesmerised by the hint of rules and progress hinted at in successional patterns. Theories of successional change were developed primarily for plant communities. The sequence of plants is a major factor in creating landscape, so succession may be important in determining plant biodiversity and, in turn, the diversity of wildlife exploiting the plants and the physical habitat they provide. Original theories saw successional change as a predicable drive towards a final ecological superorganism, the species of plants so integrated that characteristics of each community were greater than the sum of the parts, the result an inevitable **climax community**, perhaps fine-tuned to maximise efficient energy transfer. This vision faltered faced with successions leading to multiple outcomes, none conspicuously a climax, successions that appeared to reverse and no evidence for increased energy efficiency. Succession is now widely regarded as the result of individual species' life histories and attributes, each living its own life and the consistent patterns arising because of the same main environmental pressures acting on the same basic set of life forms. The attributes of plants, interactions with and changes to their environment and interactions with herbivores and diseases that exploit the plants in turn drive the process. Succession arises from the intimate ecology of physiology and life histories of species interacting with their local environment. Five processes interact: disturbance that creates an opening, immigration, establishment, competition and reaction as species alter their habitat. Important species attributes include dynamics (maximum growth rates, size, longevity and establishment) and physiology (requirements for or tolerance of light, water and nutrients).

Succession is a dynamic process and diversity can increase or decrease. Succession on virgin habitat such as a volcanic island or newly exposed glacial morraine is called primary succession. Secondary succession occurs on terrain stripped of plants but retaining previous biological influences such as soil conditioning and seed banks. Succession can be categorised by the driving forces of change. Change in response to external influences (e.g. climate change) is **allogenic**. If driven by the organisms themselves, altering their habitat succession is **autogenic**. Degradative succession occurs on organic debris such as a cow pat or dead leaf, the end result being the habitat's destruction.

Succession may be yet another ecological process that is nothing more than a statistical phenomenon that just happens to involve wildlife. Successional patterns can be modelled using nothing more than the probabilities that a species will arrive or be lost, regardless of the previous history of the site or present inhabitants. Species come and go at random.

Huston (1994: 270) comments: 'Succession is the key to understanding the regulation of nearly all aspects of biodiversity or ecological time series'. Meanwhile, Rosenzweig (1995: 63) frets: 'We have to ask the plant ecologists to return to their succession data . . . and tell us whether a consistent pattern exists'.

Interactions between species

Interactions between species can promote or destroy local diversity in the short term but are a major force for evolutionary diversification. Interactions between species fall into four broad categories:

- *Competition*: species contest a resource. Where the resource is limited inferior competitors may be driven to extinction. Competition is often depicted as detrimental to all species involved, but is often asymmetrical, with one suffering, the other unaffected.
- *Exploitation*: one species exploits another as a food source. Exploitative interactions include predation, parasitism, disease and herbivory. The victim loses, the exploiter gains. Diversity is decreased if exploiters wipe out victims or promoted if the monopoly of a few dominant competitors is prevented
- *Mutualisms*: both species benefit from the presence of the other and their interaction. Such relationships are often tightly forged, by coevolution, into a mutual dependency. Mutualisms promote diversity by opening up new ecological opportunities.
- *Engineering*: interactions due to the activities of one species benefiting others but with no reciprocal gain and typically not tightly coupled by coevolution. Ecological engineering may be physical (e.g. beaver dams) or functional (e.g. nutrient cycling).

Competition

Competition between species occurs when they use a resource in such short supply that there is not enough to sustain their potential reproduction, growth, abundance or distribution. Competition was regarded as the major process regulating local diversity. Species exactly overlapping in their exploitation of a resource could not coexist indefinitely. One wins, others lose. Competition could exclude species, lowering local diversity. However, combined with disturbance and patchiness, a mosaic of competitive interactions could result in a variety of outcomes, maintaining diversity. Longer term evolutionary pressures would select for species adapted to avoid competition by niche differentiation (living in a sufficiently different way to avoid competition), character displacement (anatomical differences), niche shifts (short-term modification of niche to avoid competition) and com-plementarity (species sharing a resource would use it at different times or places). A species cannot be good at everything and there is a strong trade-off between improvement of core abilities and evolutionary abandonment of what is done poorly. Competition provides a selection pressure with species diversifying by refinement of different abilities simply to stay in the evolutionary game. Competition theory suggested that there were limits to how similar species could be and therefore a limit to how many species could pack into an ecosystem but golden rules proved elusive. Existing niche differentiation might show that competition had been an important evolutionary pressure in the past but was not active nowadays, the species and diversity patterns ghosts of competition past. Re-evaluation of competition's role has recognised its importance but as only one of several interactions (see Figure 2.7).

Figure 2.7 Competition. Two beetle species feed on the same host plant. Competition almost wipes out one in the short term, threatening local diversity, but evolutionary diversification due to separation of feeding sites consolidates diversity in the longer term

Competition in evolutionary time is also a mixed force. Avoidance of competition by differentiation, complementary rather than overlapping exploitation of resources and exploitation of entirely new resources are powerful forces for speciation. However, the eclipse of some higher taxonomic groupings (e.g. Palaeozoic squid) ousted as dominant marine predators by the rise of the fish represents a loss of diversity.

Exploitation

Competition's fall from pre-eminence as the process determining local diversity was in part due to increased understanding of the importance of **exploitative interactions**, especially predation. Predation and competition were often cited as if mutually exclusive rivals but they have very similar effects. Exploitation can increase or decrease local diversity and is a force for evolutionary diversification. The difference lies in the structure of the interaction. Competition for food involves species feeding at the same trophic level, exploitative interactions involve links between different levels (eating your competitor could count as either). The impact of exploiters varies with the intensity of interaction, their behaviours and synergy with competitive processes. Generalist predators maintain diversity by eating anything they find in proportion to its availability or by homing in on the most abundant prey by switching their hunting activity to concentrate on it and massing where the prey is concentrated. Either strategy inflicts most damage on prey species that potentially monopolise a habitat by out-competing other species. Specialist predators may have the same effect if they specialise on would-be monopolisers but can reduce diversity if their target is a rare taxa. Exploitation works like physical disturbance. Too little allows a few species to take over, too much reduces all prey. Intermediate levels generally promote local diversity (the **intermediate predation hypothesis**). The role of predators prompted the idea of **keystone species**, which by their presence or absence define local community diversity. The keystone idea is now extended to include other roles such as keystone engineers or mutualists.

Mutualisms

Natural historians recognised the importance of **mutualisms**, perhaps as evidence of a divinely created, harmonious nature, until overthrown by post-Darwinian evolutionary theory with its imagery of contest and struggle. Mutualism became the forgotten triplet of competition and exploitation but the importance of mutually beneficial links is fundamental to ecological diversity and planetary health. Mutualisms are defined by benefit to both species, though the benefits could be seen as reciprocal selfishness rather than benign altruism. Mutualisms are typically cemented by specific behavioural or physical ties forged by co-evolution between the participants. These ties can be so specialised that neither partner can survive without the other and no alternative species can step in should one partner become extinct. Mutualisms fall into four categories. **Trophic mutualisms** are built on the exchange of food. Most often the two partners provide nutrients that the other cannot obtain itself or in sufficient quantities. Examples include coral polyp/algae, forest trees/fungal mycorrhizae and plants/nitrogen fixing bacteria. **Pollination mutualisms** and **seed dispersal mutualisms** involve plants and animals. In return for bringing pollen to a flower or transporting away seeds, the animals are rewarded with food such as nectar or pollen. **Protection mutualisms** comprise interactions in which at least one partner benefits from reduced exploitation or competition, e.g. ants tending aphids repel parasites and predators, in return receiving honey dew, while the cleaner fish ridding other fish of parasites receives a food reward.

The most biodiverse ecosystems on Earth, tropical rainforests and coral reefs, are founded on trophic mutualisms. Mutualisms open up new niches. Even species without direct mutualists have indirect benefactors. In a food chain linking plant, herbivore, predator and parasite, the plant and predator are mutualists, with the plant hosting the predator's food, the predator preventing its prey from overexploiting the plant. Similarly the herbivore and predator's parasite are mutualists. Mutualists are an additional resource prompting competition as well as cheat strategies. The image of nature conjured up by the importance and variety of mutualisms has been described as green in root and flower.

Engineering

The importance of **ecological engineers** is a recent insight, born of the increasing integration of ecosystem ecology (concerned with energy and material fluxes) and community ecology (founded on species). Living species drive abiotic processes and engineer their environment. The activities of some ecological engineers could be described as mutualistic but are different because of the lack of close evolutionary ties. In a classic example of engineering, the beaver, constructing a pond for its own benefit, inadvertently provides a habitat for many other species. The beaver's actions are vital but not co-evolved with the beneficiaries. Engineering therefore includes many interactions where one participant benefits, the other is unaffected. Engineers' impacts work by the impact on either biogeochemical cycles or the habitat structure. Some **structural engineers** create physical structures out of their own bodies (e.g. coral and trees), so-called **autogenic engineering**. Others create alter existing structures (e.g. beavers and woodpeckers), so-called **allogenic engineering**. **Biogeochemical engineers** alter chemical cycles and nutrient availability, for example bioturbation of sediments altering nutrient (resource) and oxygen (regulation) gradients.

Reawakening interest in mutualisms and the importance of engineers driving ecosystem function have altered our understanding of the role of species interactions as a source of biodiversity. Competitive and exploitative interactions are very important, creating local variations and evolutionary escalation. Besides the direct impact of exploiters on victims and competitors on each other, indirect effects ripple through ecosystems. Many competitors or victims may escape single, strong interactions but are buffeted by many diffuse conflicts. The parasites of a herbivore are the plants' mutualists. Two competitors that share a predator magnify predation on each other since their enemy has a larger total pool of prey. Two competitors with different predators benefit from the activities of each other's enemies. Meanwhile ecological engineers riddle their habitats with geochemical variation and physical structure. Evolution multiplies these intimate links, particularly host–parasite systems as host speciation spawns parasite co-speciation. This intimate tangle, populated by mutualists and driven by ecological engineers, has challenged the mechanical, reductionist ecology of biodiversity. Ecosystems may be examples of **complex adaptive systems**, defined by their diversity of components, complexity derived from simple rules generating varied patterns, with energy and materials in constant flux and the ability to adapt and evolve. The sense of patterns emerging that are more than or at least different from the sum of their parts shows the wheel of scientific fortune changing. Species interactions are vital to the generation and maintenance of biodiversity. So, biodiversity begets biodiversity after all (see Figure 2.8).

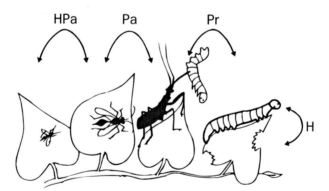

Figure 2.8 Trophic interactions. Herbivory, H; predation, Pr; parasitism, Pa; Hyper-parasitism (parasite of a parasite), Hpa. In addition the assassin bug predator of the caterpillar is a mutualist of the plant, killing herbivores. The hyper-parasite a mutualist of the assassin bug, killing parasites so restricting their future impact on the bug, and therefore is also a mutualist of the plant

Disturbance

Natural disturbance is an important factor creating or destroying local diversity. Disturbance takes many forms and acts at many scales. As an intimate ecological factor, disturbance includes physical abiotic impacts such as fire, flood, storm, landslide and treefall. Disturbance is defined as a discrete event, a sudden physical destruction, causing mortality faster than populations or cover can be replaced. The loss frees up resources which can be exploited by survivors and colonists. Disturbance is different from continuous stress which slows growth or rare, massive catastrophes such as meteor impacts. Local physical disturbance is the ecological process outlined here, though predation and disease are also a form of disturbance, but the result of biotic interactions.

The impact of any disruption varies with five factors: disturbance frequency (number and predictability), intensity, synchrony (one large disturbance versus many, smaller out of synch patches plus timing relative to victims' life histories), size of gap that results, and effects on resources. The disturbance alters biodiversity by immediate deaths, changes to resource availability (amount and type), physical alteration of the habitat and evolutionary adaptation.

The effect of disturbance is inextricably linked to successional, competitive interactions and habitat patchiness. Disturbance disrupts inevitable outcomes, particularly monopolisation of habitat and resources by a few superior competitors. The disturbance may keep populations so low that resources are never in short supply and competition does not occur. Or disturbance can reset and redirect the struggle, creating diverse outcomes, especially if separate patches are perturbed at different times and intensities, resulting in a mosaic of communities. The disturbance may alter a patch's environment so much that it is no longer hospitable for previous occupants. Disturbance plays a vital role in conspiring with deterministic forces, such as competition, and physical structure, such as heterogeneity, to create dynamic patterns of biodiversity (see Plate 12).

Disturbance can also decrease biodiversity, if too frequent, severe or extensive to cause extinctions and inhibit recovery. Diversity is greatest at intermediate levels of disturbance, a general result encapsulated in the **intermediate disturbance hypothesis**. Too much disturbance wipes out species, too little allows monopoly. Defining intermediate is awkward and dangerously circular: intermediate disturbance promotes diversity because

Plate 12 Disturbance and succession. Brown *Fucus* seaweeds dominate the middle levels of this rocky shore but patches have been ripped out by disturbance from waves. These openings have been colonised by fast growing green algae so that local diversity increases. A succession of seaweeds will invade, determined by individual attributes of colonisation, competitive ability and resistance to grazers.

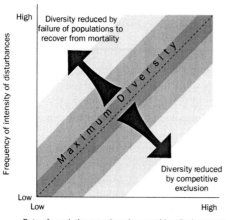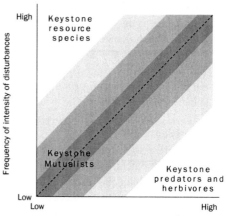

Figure 2.9 Huston's 1994 synthesis of disturbance, population growth and interactions as determinants of local diversity. The left graph shows diversity reduced if either growth and competition or disturbance is high, without any compensating influence. The right graph shows the importance of different keystone interactions under three disturbance/population growth regimes

Source: Redrawn from Houston (1994).

it is a level of disturbance at which diversity is maximised. It is better to consider disturbance and competitive population growth rate processes simultaneously. The two have been formally linked in the **dynamic equilibrium model** of local diversity. The link highlights intermediate disturbance promoting diversity when growth rates and competition are high. High growth rates as a result of productivity promote diversity by aiding recovery from disturbance. The model suggests disturbance effects vary with trophic level. High level taxa have smaller populations, which are more vulnerable to extinction and require greater energy to recover. An increase in disturbance will endanger higher level diversity more than that of lower trophic levels, especially when productivity is low. At high productivities but low disturbance, biotic impacts of keystone predators and herbivores become important as a means of breaking up low diversity monopolies. At low productivities but high disturbance, keystone resource species accelerating resource supply are important. At low productivities and disturbance, keystone mutualists are important (see Figure 2.9).

Dispersal

Diversity varies with species' abilities to reach a site which in turn depends on attributes and behaviour. Dispersal, the movement or transport of individuals, can be very conspicuous whether it is dandelion seeds on a summer's breeze or herds of migrating wildebeest. Because dispersal is awkward to measure or model and the fate of those that do not complete their journey often unknown, its role was poorly researched. Biodiversity deals in variety, populations, communities and ecosystems but it is individuals that disperse. Individual based models, taking into account that all individuals and their fates are different, have advanced our understanding of dispersal's consequences for local diversity. At the opposite extreme, large-scale **supply side ecology**, the rate and timing of recruitment, particularly of dispersing juvenile stages of marine life swirling in gigantic oceanic currents, helps to explain regional and local diversity.

Dispersal is commonly divided between **migration**, the active, mass movements of whole populations, and **dispersal**, the active or passive transport of individuals. The individuals in question may be adults, juveniles, eggs, spores or bits capable of vegetative growth. The precise mechanisms, timing and rates are very species specific. Marine plankton may disperse at the mercy of the tides. Some plant seeds disperse in tightly co-evolved mutualisms with one species of animal carrier. All species have some power of dispersal, even those in uniform, stable habitats. Different types of dispersal create different patterns. **Jump dispersal** over great distances and barriers may found regional populations. **Metapopulation** species rely on dispersal to maintain their scatter of discrete colonies. **Diffusion**, the gradual movement and expansion of range, will mesh with habitat heterogeneity, resulting in an unevenly spread population. Even in a homogenous terrain, diffusion can create patterns. The spread of populations, oscillating through peaks and troughs of abundance typical of many species, will meet like waves on the surface of water. As high and low populations of interacting species ripple across the landscape, complex patterns emerge.

Supply side ecology, the rate and pattern of arrival, can be a dominant influence. Inconsistencies in the diversity of many coral reefs and marine benthic communities have shown that variable recruitment, driven by the rate and timing of dispersal, causes local populations occasionally to be deluged by arrivals, whilst other areas receive none. These occasional events swamp local interactions that regulate diversity in between. There are lessons for conservation. The plagues of the crown of thorns starfish (*Ancanthaster planci*), famed for the havoc they wreck on corals, may outbreak due to occasional unpredictable mass arrival of larvae.

Ultimately the dispersal of individuals builds the ranges within which evolutionary events (such as speciation) and ecological primary factors (such as vicariation) work. From time and space to individuals and back again, the ecological factors creating biodiversity are an integrated loop rather than a reductionist shopping list.

Ecological renaissance: emerging general rules

Ecology has at times appeared an intractable science. Atoms are so small and galaxies so large that they allow mathematical generalisations but ecosystems and their inhabitants appeared to be messy, contingent and unpredictable. Whilst ecologists could do experiments and explain in detail what happened to one species of water bug on one marsh or tussocks of grass up on a mountain, it was difficult to extend these lessons to anywhere else. However the period since the mid-1990s has seen significant developments which hint at unifying principles and general ecological rules.

The relationship between size and **metabolic rate** has been a key area. Metabolic rate reflects many key aspects of a species' life such as growth and reproduction, from bacteria to blue whales. Metabolic rate varies consistently with body size so that smaller organisms use more energy per gram of body weight than bigger species. Work on forest trees has shown consistent relationships between size and abundance so that for every tenfold increase in tree trunk diameter, the number of trees decrease by a hundredfold, but the horde of smaller saplings uses the same total energy as the smaller number of larger trees. Each size class uses the same amount. Similar patterns occur in animal communities. These consistent patterns suggest a template for the distribution of resources and, because of the relationship between metabolism and body size, the diversity of life.

A second approach is **macroecology**, which recognises that the local species richness (those bugs in the pond or tussocks on the hill) is influenced by distributions and abundance at larger spatial and temporal scales. Macroecology has highlighted several key patterns where the influence of the large scale on the local is apparent. First, there are the familiar relationships between **species richness** and area, isolation latitude and altitude. Second, consistent **range size** patterns such as species at higher latitudes have wider ranges and species with wide ranges occur at more local sites than species with narrow ranges geographical. **Abundance patterns** include individually abundant species having wider ranges than rare species and consistent patterns of a few abundant and many rare species in most communities. Finally there are **body size relationships,** such as the greater number and higher populations of smaller species compared to larger ones. These large scale phenomena, broadly consistent across taxa, habitats and time, have been explained based on models using the relationship between energy and biomass and the constraints these inevitably impose on species.

A major ecological conundrum may also be near resolution – that complexity (e.g. more species, more interactions) fuels instability (e.g. extinctions). The revelation that complexity begat instability came from models of food webs in the 1970s and did much to shatter the early twentieth century view that greater ecological complexity reinforced stability and that natural systems existed in some sort of equilibrium. However, the result is also at odds with the conspicuous complexity of many ecosystems such as tropical forests and coral reefs, which do not spontaneously collapse. New models have built in more realistic species dynamics and results suggest that many weak interactions result in greater stability. Once again, all those species matter.

The final and boldest development has been a unified theory of biodiversity and biogeography (biodiversity in this case defined strictly as the richness and abundance of species). This model unifies the competing visions of ecological communities as regulated either by strong interactions between species fighting to occupy a limited number of niches or as looser assemblages thrown together by dispersal and extinctions. The theory makes the assumption that all individuals in a community with the same broad trophic role, e.g. all predators or all herbivores, are equal in terms of competitive ability and chance of survival, although different species vary, for example in their likelihood of colonising. The model then assumes that there is a limited space/resource to fill up, that new individuals can establish only if others are lost and that individuals survive or are lost at random as space becomes available. The underlying maths is fearsome but the model produces strikingly realistic predictions of species richness and abundance across a range of ecosystems and a single parameter, based on two variables (total number of individuals in the regional community in question and the rate of speciation), is sufficient to define the outcome for each community.

The evolution of biodiversity

The ecology describes patterns of and provides explanations for the biodiversity of extant ecosystems. Ecological processes also have evolutionary consequences. They interact with genetic diversity via adaptation, microevolution and speciation. The day-to-day dramas played out on the ecological stage are the driving force for the multiplication of species in evolutionary time. The environment provides continual pressures to diversify via adaptation, innovation and exploitation of new ways of life. Taxonomic biodiversity is the most conspicuous result of ecological and evolutionary processes driving the multiplication of species.

The multiplication of species

Diversification of genes: environmental forces

New species arise from two sources. An individual lineage may change over time so that subsequent generations are sufficiently different from ancestral types that they are classified as a separate species. This is **anagenesis** or **chronospeciation**. There is no net gain of species. Speciation that spawns a net gain in numbers is **cladogenesis**, whether it be splitting of the ancestral line into two (or more) new species or budding of new species from an ancestral lineage that also survives. These processes can be gradual, the accumulation of inherited genetic differences selected by ecological processes, or sudden, perhaps random, with no clear adaptational benefit – Raup's survival of the luckiest.

Taxonomic biodiversity is increased by cladogenesis so it is important to understand why lineages speciate. The genetic variation in populations that underpins all of biodiversity and is central to evolutionary processes is almost unavoidable. Genetic diversity within a species is created both by the environment, as local populations adapt to variations between local conditions, and by processes within the genome that would occur even in a homogenous environment, e.g. mutation. The drive to genetic diversity is irresistible.

Some genetic diversity can be categorised as **adaptational**. Genetic differences reflect variations in local conditions with selection of genotypes beneficial in the prevailing conditions and loss of the sub-optimal. The genetic differences may reflect tolerances to different conditions or create a functional diversity so that individuals in a population can exploit varied opportunities, but perfect adaptation is probably impossible, not least because environments constantly alter. Adaptational diversity may create **key innovations**, allowing exploitation of lifestyles new to that taxa or novel to any life. Key innovations are one evolutionary factor unleashing adaptive radiations, resulting in geologically rapid diversification of a lineage. Innovations may be utterly new. They also include **pre-adaptations**, a character allowing ecological or evolutionary exploitation of a new lifestyle whilst essentially carrying out its original function, and **exaptations**, characteristics again allowing ecological or evolutionary expansion, but the character is used in a different way from its original purpose, fortuitously effective in a new role.

Major periods of innovation, defined by their global extent, diversity of innovations and new ecosystems may spring from revolutions in global biogeochemical cycles. The two main innovation revolutions since the Cambrian (one the Cambrian marine diversification, the other the Mesozoic marine modernisation) coincide with massive **biosphere** changes such as volcanism, warming and sea level changes which could alter nutrient supply. Increased supplies of nutrients and energy opened up opportunities and once evolution and speciation accelerated, a positive feedback of arms races, competition and mutualisms would snowball.

Genetic and resultant evolutionary biodiversity can also arise without any adaptational context. The exchange of genetic material through an interbreeding population results in **genetic drift**. Genetic drift can be a potent force in small populations where the original genetic diversity may be limited to start with, a biased sample of the total variety in a larger population. In a small population a few matings can result in drift rippling through subsequent generations. Small populations prone to such effects can be created in many ways: **founder populations** of a few individuals colonising new habitat; isolated remnants marooned in refugia; **bottlenecks** created when populations crash; metapopulations where the total population may be high but is scattered between separate patches that do not readily interbreed.

Diversification of genes: genetic systems

Just as the environment forges genetic variation within populations, so processes within the genome itself create and maintain diversity. **Mutations** are any changes in the genome, whether one **nucleotide** or larger sequences. Types of mutation are unpredictable, uncontrolled and independent of the natural environment, though rates may vary. Mutations include errors in **DNA (deoxyribonucleic acid)** replication, insertion of transposed sections and rearrangement of chromosomes via fusions and inversions. The rate of changes to Eukaryote nucleotides has been estimated as 4×10^{-9} per base pair per generation. Since eukaryotes typically harbour billions of base pairs, every individual carries some mutations. We are all mutants. So mutation will create genetic variation even within the least genetically diverse taxon, a source of hope for genetic recovery for some endangered species. Genetic drift creates diversity within a freely mating population. Although the frequency with which individual **alleles** are transmitted to offspring may be broadly predictable (based on their frequency in the parent population and the rules which govern their sharing out between gametes), only a minute fraction of gametes actually contribute to the next generation at fertilisation. The alleles in the billions that do not make it are winnowed. The impact of drift depends on population size, and are magnified in small populations. Drift effects can be substantial in large populations, for example species with breeding opportunities monopolised by a few individuals such as red deer stags coveting harems. The limited number of individuals contributing to the genetic diversity creates a small **effective population size**.

The distribution of individual alleles on chromosomes can also be rearranged by **recombination**. Eukaryote chromosomes exchange sections so that alleles that would otherwise stay linked together are combined into new sets. The total genetic diversity at the allele level does not increase but the permutations do. Recombination is most effective in **outbreeding** populations. Self-fertile taxa end up recombining with themselves. Prokaryotes have mechanisms that achieve broadly similar remixing of alleles.

Selection and maintenance of genetic diversity

Environmental and genome processes combine to produce genetic diversity within populations; the genetic foundation of each individual is tested by **selection**. Selection occurs within ecological time and forges a lineage through evolutionary time.

Selection acts directly on the phenotype. The genotype is affected indirectly. **Directional selection** favours individuals with an advantageous characteristic, selecting the genotype responsible and reducing genetic diversity by elimination of alternatives. Since many general phenotype characteristics are a synergy of complementary features not governed by one gene, directional selection effectively selects sets of genes. **Stabilising selection** favours intermediate characteristics and extremes are winnowed out. Stabilising selection fixes combinations of alleles that build the intermediate character rather than variations within each allele so again genetic diversity is lost. **Disruptive selection** favours different extremes; the intermediates are at a disadvantage and maintain the variation in alleles that cause the switch between extremes, not different alleles themselves. So selection tends to decrease genetic diversity. However, fluctuations through space and time alter selection pressures, even reversing them, providing an important brake to such losses. **Genetic dispersal** and **gene flow** act to homogenise genetic diversity within a population. Highly mobile species show less variation than the more sedentary. The dispersal characteristics of a species are important. Some species known to be poor dispersers nonetheless show little genetic variation throughout their range. This may be because they have only recently expanded into the range, the homogeneity an echo of their close-knit origins. Species living as a metapopulation can show marked losses of genome diversity if extinction of patches wipes out unique variations. The toll from selection and drift is countered by imports from flow. The outcome teeters, critically dependent on how far individuals move between birth and mating versus the strength of selection. There can be a critical distance below which flow can compensate for any effects of selection. The conservation of small, fragmented populations often depends on understanding such pressures. In summary, mutation creates genetic diversity, whilst selection and drift destroy.

Breeding systems affect genetic diversity. **Inbreeding** allows disadvantageous, often recessive alleles expression. Enfeebled offspring show **inbreeding depression**. If they breed in turn, the degradation may snowball into a genetic meltdown. This is a particular threat if a once large population is suddenly reduced in size. Genetic meltdown becomes the final straw after the first strike that reduced abundance. Small founder populations which may be **homozygous** to start with so that any deleterious alleles have been weeded out already are in less danger. Outbreeding generally suppresses the worst impacts of disadvantageous alleles. **Outbreeding depression** can occur via immigration of individuals poorly adapted to local conditions which mate with well-adapted locals. There is a balance: mate with those too close and risk inbreeding, mate with those from too far and risk outbreeding depression.

Given the environmental and genome promotion of genetic diversity, it is not surprising that there is no evidence of limits to local, small-scale selection for want of genetic variation. We do not know if there are limits to major evolutionary changes, for instance radical rearrangement of body plans. These may be prevented by lack of genetic foundation. Perhaps the genes do exist but developmental rules halt any such revolutions. Even without the capacity for unfettered creativity, the environment and genes have conspired to create a wealth of species.

Speciation

Speciation is the creation of new species. Whilst commonly pictured as a diversification or radiation of a lineage, speciation requires mechanisms both to create and then maintain the diversification into cohesive species. Only then can a genotype establish, its individuals isolated from interbreeding with other species. The familiar images of radiating, diversifying lines might be better thought of as collapsed spokes from a wider fan of individuals that once freely interbred.

The threshold from variation within a species (adaptation, microevolution) to speciation and macroevolution is a barrier. The very same processes creating genetic diversity within a species also maintain cohesion. Gene flow may smother variation; unusual genotypes may be less fertile or selected out by the environment or ignored by would-be mates. Selection may hone well-adapted species, marooned on peaks of near perfection, from which any evolutionary departure is difficult.

Mechanisms are needed to push genetic diversity over the speciation threshold. There are four main modes of speciation:

- *Allopatric:* geographical separation of a previously continuous population.
- *Peripatric:* a variant of **allopatric** involving isolation of a peripheral population.
- *Parapatric:* speciation in adjacent populations stretched over an environmental gradient.
- *Sympatric:* no separation during speciation.

Allopatric speciation

This requires the separation of populations either by creation of a physical barrier, extinction of the middle range of a population or jump dispersal across an existing barrier prior to speciation. The barrier prevents gene flow and the populations' genomes diverge. The divergence can be adaptive if environments differ either side of the barrier or random due to mutations, recombinations and chance losses. If the diverging populations rejoin, they may still interbreed and coalesce, maintain the divergence but with a hybrid zone in between, or not interbreed at all, the latter representing full speciation. This reproductive isolation can be maintained by genetic incompatibility, discrimination in choice of mate or the physical impossibility of mating (see Figure 2.10).

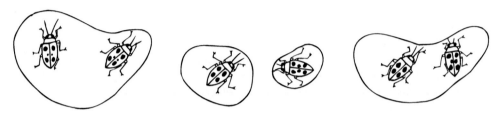

Figure 2.10 Allopatric speciation. A united population is split, in this case by a physical barrier of rising sea levels. The evolutionary history of the isolated populations differ, to the extent that once reunited they no longer successfully interbreed. Two new species exist where there had been one

Peripatric speciation

This involves isolation of a small peripheral population from the main range. Otherwise the process of geographical isolation followed by divergence are the same as allopatric speciation. The small size of peripheral populations creates specific features. Small numbers of founders inevitably contain only limited, biased subset of the original species' gene pool. This provides a head start to divergence, magnified by inbreeding and homozygosity, allowing expression of unusual characters previously suppressed. The cause of isolation may add impetus. Founders colonising an island may have abundant resources and little competition. Alternatively peripatric speciation in collapsed remnants of larger populations can face genetic bottlenecks. These **founder effects** emphasise the vagaries but importance of **chance events** and genetic loss. Alternatively, adaptations may flourish in new or empty niches. Old genetic complexes are shaken up, innovations and radiations are marked. This adaptive, creative result is called the **founder flush**.

Parapatric speciation

Variation throughout the range of a species and resulting local adaptation will create genetic gradients. Once such a cline starts, gene flow across the gradient may decrease, especially if dispersal is poor, there is selection against hybrids and increasingly pure types straying into the wrong side of the cline. The more pronounced that the differences become, the more the trend is reinforced. A true hybrid zone will develop and, if reproductive isolation becomes established, eventually will disappear leaving two adjacent species. Distance over the range of the populations is the creative force, since genetic variation driven by genome mechanisms alone can create clines even across a homogenous habitat. Geographical and ecological disruption can aid and abet progress.

Sympatric speciation

This involves no spatial separation of population. Separate, cohesive genotypes are established and maintained whilst in contact with one another. Models of sympatric speciation have relied heavily on selection.

- **Disruptive selection** by selection of different phenotypes is controlled by one gene and elimination of intermediates. The selection of phenotypes may be due to ecological interactions such as predation. Once different phenotypes are established, natural selection will encourage reproductive isolation through habitat selection or positive assortive mating (different phenotypes positively choose to mate with their own kind). Sympatric speciation linked to habitat selection has been suggested as a major cause of insect diversity (see Figure 2.11).
- **Competitive speciation** is a variant of disruptive selection. Competition for resources within a species selects for different phenotypes able to avoid intense competition whilst intermediate types are lost.
- **Polyploidy** relies on the multiplication of whole sets of chromosomes so that off-spring have double (or more) the normal complement. Gametes of polyploids typically contain more than the normal one-half set of chromosomes, so if fertilised (whether by another polyploid or not) the offspring has a different number of chromosomes. Polyploidy relies on abnormal cell division to create the overloaded gametes. **Auto-polyploidy** occurs by fertilisation between polyploids within a species. The resulting offspring often suffer low fertility, their mass of chromosome literally tangled during cell division to create gametes but vegetative reproduction can be strong. The potato is a good example. **Allopolyploidy** relies on hybridisation between different species of polyploids. Offspring are usually fully fertile, gamete production smoothed as the chromosome contributions of each parent separate out with their own kind, avoiding tangles. Many wheat varieties are allopolyploids. Polyploids are often bigger and more productive than their forebears. Polyploidy is rare in animals but a major speciation mechanism in plants, with over 43 per cent of dicotyledon and 58 per cent of mono-cotyledon species polyploid.

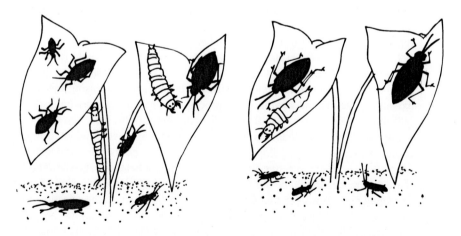

Figure 2.11 Sympatric speciation by disruptive selection. A species of beetle, which shows wide variation in body size, shares part of its range (on the plant) with a predator. The predator can overpower small but not large beetles. Eventually large and small beetle populations exist as a result of predator selection. If differences between the beetle populations are consolidated so that they no longer interbreed successfully two new species exist where there had been one

Other potential sympatric mechanisms include chromosomal rearrangements that result in reduced fertility of **heterozygotes**. In the meantime **homozygotes**, potentially displaying distinctive features, can establish populations which become cohesive, much like disruptive selection. This is called **stasipatric speciation**. **Macrogenesis**, the dramatic mutation to produce a hopeful monster, is a possibility, demonstrable in some laboratory populations but with little support in nature.

The relative importance of all mechanisms is still hotly debated. Examples occur of all types, their precise role dependent on the taxa in question.

Speciation in action

Speciation, especially the evolutionary consequences on a geological time scale, are a culturally familiar idea in the developed world, for example the diversity of the iconic Darwin's finches on the Galapagos Islands. However, are there examples of speciation taking place before our eyes to back up and test the different models and hypotheses? Yes, in particular amongst insects, which are both familiar and have fast enough generation times to study.

A neat example of peripatric events is shown by the populations of the mosquito *Culex pipiens* in the tunnels of the London Underground. The subterranean populations are genetically distinct from their above-ground relatives. Populations within individual tunnels are also genetically distinctive: the limited genetic diversity within the tunnel populations is suggestive of founder effect bottlenecks providing only a limited gene pool and that the foundation of new colonies is a rare event. Whilst mosquitoes from different Underground populations will still interbreed with one another if brought together, they very rarely will do so with above-ground mosquitoes. The London Underground system was begun only in 1863. This is incipient speciation within a few human lifetimes.

A similar example occurs in the natural tunnels left behind by lava flows in Hawaii. Within these caverns live troglodyte plant hoppers of the genus *Oliarus*. One species, *Oliarus polyphemus*, has been studied in detail. Plant hoppers from populations in adjacent tunnels look the same but have distinctive courtship songs, whilst those within the same tunnel sing to the same music sheet. They are reproductively isolated not just physically, tunnel to tunnel, but also behaviourally, by song. *Oliarus polyphemus* is regarded as a complex of incipient species.

Insects also provide examples of sympatric speciation. The wild fruitflies *Drosophila melanogaster* of 'Evolution Canyon' on Mount Carmel in Israel are a fine example, not least because so much laboratory work has been carried out using this very species to examine genetics and evolutionary processes. The canyon has opposite slopes only 100–400 metres apart, well within the dispersal range of the flies. The south-facing slopes are hot and dry, and conditions fluctuate creating a more stressful environment than the cooler, wetter north facing slopes. Flies from either slope show marked, genetically based adaptations to their respective slopes, for example for heat tolerance or drought stress, but these differences should be obliterated by dispersal of flies and mixing of genes between the populations. However, flies from either slope show very strong selection for mates from their own slope. Again, incipient sexual isolation although the flies are not physically separated, unlike the Underground's mosquitoes or Hawaii's *Oliarus* bugs.

What do all those species do? Part I: biodiversity and ecosystem function

By the late 1980s ecology was bogged down in a circular debate; does greater complexity (for example higher species richness) create greater stability or vice versa? Ecologists concerned primarily with species richness and abundance within communities operated an almost separate discipline to those interested in ecosystems and the flow of energy and nutrients. The role of species in driving the Earth's geochemical and energy cycles went largely ignored. However, in 1991 a meeting of environmental scientists in Bayreuth, Germany, responded to this growing fault line, addressing two key questions. First does the richness of life matter for ecosystem processes? Second, how might the stability of ecosystems, both their resistance to change and their resilience to recover, depend on species richness? The resulting book *Biodiversity and Ecosystem Function* (Schulze and Mooney, 1993) fuelled a renaissance of interest in the relationship between ecological richness and ecosystems processes (biodiversity and ecosystem functions, in the new jargon). The field of biodiversity-function research blossomed in the 1990s driven in part by the growing awareness of the value of ecosystem services and the threat to us all as habitats degrade and species are lost. Ground-breaking research produced tantalising results but, like all new science, acrimony. Science journalists have spotlighted the tensions with headlines such as 'rift over biodiversity divides ecologists'. Nonetheless general patterns are emerging.

Initial studies tried to conceptualise the ways in which ecosystem function might vary with species richness. There are several models, all commonly describing how the effectiveness of ecosystem function varies as species richness declines.

- *Linear models:* ecosystem function degrades in proportion to loss of species. In effect every species matters, more or less equally.
- *Redundancy models:* as species richness declines, the ecosystem function remains largely unimpaired as other species in the habitat take up the functional roles of those that are lost. Species can substitute for one another. There are in effect spare species in the habitat. Ecosystem function collapses sharply once a minimum threshold of species richness has been breached.
- *Singular models:* these models regard all species as making a unique contribution. The rivet model uses the analogy of rivets popping out of a machine: lose a few and not much changes, but once too many are lost, a bit of the machine falls off. In natural systems the equivalent is that several species may be lost with little impact until enough are lost to cause a sudden loss of ecosystem function. Functional effectiveness then stabilises until the next time when enough remaining species have been lost. In the keystone model, ecosystem function depends on one vital species to drive the process. Once the keystone is lost, ecosystem function degrades markedly whilst losses of other species has little effect.
- *Idiosyncratic models:* the complexity, contingency and context of ecological inter-actions might make the outcome of species losses unpredictable. Ecosystem function may degrade or improve, depending on which species are lost, in which combinations and when and where.

Whilst these representations of the relationship are helpful, they do not explain how the processes might work. Three main hypotheses have emerged:

- *Sampling effect:* this explanation suggests that the more species there are in a habitat, out of those locally available, then it is more likely that the most functionally effective of the local species pool will be included. So, on average, the more species there are, then the more likely it is that the most productive or most effective nutrient conservers are present. However, a monoculture of the most effective species would be every bit as effective so this hypothesis simply recognises the results of sampling any array of items, rather than suggesting that diversity itself creates any special bonus.

- *Species complementarity or niche differentiation models:* species vary in their needs, the resources they use and their precise ways of life. The more species there are in an ecosystem, the wider the range of resources that will be used and processes that will be biologically driven.
- *Positive interactions:* species may interact in ways which help one another, either through specific mutualisms in which species are adapted to benefit one another or by altering the environment in ways which accidentally facilitate survival of other species.

Some classic studies

Despite the apparent simplicity of the key question – does biodiversity affect ecosystem function? – experiments to investigate the relationship have proved very demanding of resources and a minefield for interpretation. There are now several classic studies which have produced robust results.

Ecotron, UK

Pioneering experiments, the Ecotron, from 1993 onwards using sixteen climate-controlled rooms in which miniature ecosystems could be created and species dynamics and geochemical cycles carefully monitored. Experiments could combine multitrophic communities – plants, herbivores, predators and parasites – and include both above and below soil systems. Experiments could be run over several years. The Ecotron facility was used to test the effects of varying species richness on ecosystem functions within the microcosms. Lower species richness resulted in lower overall productivity, the first experimental demonstration that for effective ecosystem functions the number of species in a habitat matters.

Cedar Creek, USA

The Cedar Creek Biodiversity experiments use outdoor plots of prairie grassland plants to examine how species richness affects plant productivity and geochemical functions such as carbon and nitrogen cycling. Individual plots can be planted with varying numbers of species, between one and twenty-four, including a variety of functional groups (e.g. nitrogen fixers) and tracked over several years as the communities and functions respond to natural variations in climate; responses of insects, diseases and invasive plants can also be monitored. Key results are that both productivity and nutrient use increase with species richness. Productivity increases not only are linked to the presence of one species (so, not the sampling effect model) but also are long-term responses to different combinations of species, in particular the variety of functional groups. Both species richness and functional types matter.

Biodepth, Europe

The Biodepth project uses a similar approach to Cedar Creek, extended across geographically diverse context. Plots of plants, varying in richness, from one to thirty-two and in functional groups, have been set up in seven countries, from Ireland to Greece and Sweden to Portugal. Ecosystem processes such as productivity, decomposition and nutrient cycling have been monitored. The key results are that above-ground biomass of plants increased with species richness and variety of functional groups but there was also variation with the precise composition of species and between sites. Species richness and variety both matter, but there is evidence of idiosyncratic responses to combinations of species and site.

All these projects are ongoing, increasingly focused on how the systems respond to environmental change, especially to climate. Whilst the debate remains fierce and the species and habitats involved vary, the general message is unmistakable; the variety and richness of life on Earth is fundamental to Earth's natural economy. Living things drive the global environment, not into mythical, idyllic balance but quite the reverse, preventing the planet closing down.

What do all those species do? Part II: ecosystem integrity and ecosystem health

The significance of biodiversity for our well-being combined with the wide conceptual horizons of biodiversity, connecting the biological sciences to economics, politics and culture, have encouraged ecologists to develop more holistic paradigms of natural systems, combining the species, the ecosystem functions and how these affect humanity. One approach has been the idea of **ecosystem integrity**. Put simply this is the ability of the ecosystem to support and maintain the full range of species, communities and processes expected in the natural habitats of a region. Note how this combines four elements: first, the structure and organisation (e.g. numbers of species, relative abundance); second, the natural functions (or vigour), third, the ability to resist damage or the resilience to recover and fourth, the undegraded potential to develop and adapt to change. Degradation of any of these represents a loss of integrity. This approach alters how we think about the natural world. The habitats within a nature reserve may support a wealth of species and communities and be apparently pristine, but the ecosystem may have limited opportunities to change or adapt, perhaps because we manage it to meet our targets, stifling change, or perhaps because there is not the room or resources to allow change. The notion that change is an important aspect of natural systems also challenges how we conserve systems. There is a movement to allow 'wilding' of conservation landscapes, where we do not intervene to meet chosen management priorities but let nature do whatever it will, even if the precise end result is not known.

The 1992 biodiversity convention talked of 'the need to safeguard the health and integrity of the world's ecosystems'. The very existence of something called **ecosystem health** has been controversial amongst scientists, but, like integrity, is an approach recognising the need to cross disciplines and boundaries to effectively conserve the natural world. The notion of ecosystem health recognises that ecosystems are degraded by the stresses we inflict overwhelming the natural resistance and resilience. The ecosystem becomes dysfunctional, just like we do when we are poorly, and that an approach of both prevention and cure is needed. Ecosystem health emphasises the need to define and measure loss of function, to diagnose causes and detect early warnings. The idea is an effective analogy, helping to get across the impact we are having on the natural world. Ecosystem health echoes the holistic approach of ecosystem health, incorporating the vigour, organisation and ability of ecosystems to resist and recover from damage. The idea has attracted flak, sometimes conceptual such as the implication that ecosystem may be like superorganisms, sometimes pedantic such as the suggestion that the word 'health' cannot be used like this. The key ideas of integrity and health have been taken on board. In 2004 the UK government conservation umbrella the Joint Nature Conservation Committee (JNCC) published guidelines for site and habitat assessment which recognised the need to measure both the structure and function of habitats and use these to judge their condition against their desired state, rather than just taking a snapshot of their condition, their publication popularised explicitly using the idea of ecosystem health.

As the idea of biodiversity has encouraged the natural sciences to engage with the wider world, the real problem could be that many of these useful ideas become muddled together. Ecosystem integrity emphasises the role of **natural ecological and evolutionary processes**, wherever they may lead, and the conservation of natural organisation and function. Ecosystem health emphasises the **measurement** of resilience, vigour and productivity and the need to detect and prevent their degradation. Both are allied to the concept of **sustainable development** which emphasises **economic processes** and development without compromising the future well-being of the planet, which is where the whole UN involvement in biodiversity and the Rio treaty began. They are all different in detail but all three recognise the same essentials for our future survival.

The imbalance of nature

The balance of nature has a strong hold on our imagination and conveys a cosy vision of harmony. However, nature's creative imperative is anything but balanced. The diversity of life around us, at least the most familiar aspect of species richness, can be reduced to a very simple equation. Species diversity equals speciations minus extinctions. Since there are still very many species alive on Earth, nature is profoundly imbalanced, with speciation generating a surplus over and above the ravages of extinction. Quite how big that surplus is remains one of the most difficult questions facing biologists. The diversity of life on Earth is still surprisingly poorly known. Chapter 3 provides an inventory of biodiversity, reviews progress and problems and tries to quantify the remaining surplus of nature's creativity.

Summary

- Ecological and evolutionary sciences seek to explain patterns and processes of biodiversity.
- Ecological factors controlling diversity range from large scale influences of time and geography to intimate interactions between species and species' own attributes for dispersal.
- The diversification of genetic material ultimately spawns new species.

Discussion questions

1   Why do we believe in a 'Balance of Nature'?
2   How do wild animals in your country engineer their habitat?
3   Why is the biodiversity of tropical rainforests so great?

Further reading

See also

The causes of extinction, Chapter 4 pp138–151.
Natural extinctions, Chapter 1 pp32–37.

General further reading

Barbault, R. and Sastrapradja, S.D. (1995) *Generation, maintenance and Loss of Biodiversity*. In V.H. Heywood (ed.) *Global Biodiversity Assessment*. Cambridge University Press, Cambridge.

Detailed review of ecological and evolutionary processes creating biodiversity.

Dawkins, R. (1986) *The Blind Watchmaker*. Longman, Harlow.
Richard Dawkins' tour de force explaining evolution's creative power.

Huston, M.A. (1994) *Biological Diversity: The Coexistence of Species on Changing Landscapes*. Cambridge University Press, Cambridge.

Rosenzweig, M.L. (1995) *Species Diversity in Time and Space*. Cambridge University Press, Cambridge.
Huston and Rosenzweig's books are reviews of ecological processes and how these influence biodiversity. They are also very different in emphasis, Huston concentrating on the role of disturbance and interactions, Rosenzweig on area.

Wilson, E.O. (1994) *The Diversity of Life*. Penguin, London.
Beautiful exploration of ecology and evolution and the creation of biodiversity plus human impact.

Other resources

Ecosystem function research

The BIODEPTH work can be accessed at http://www.cpb.bio.ic.ac.uk/biodpeth/
The Cedar Creek work can be accessed at http://www.cedarcreek.umn.edu/research
The ECOTRON work can be accessed at http://www.cpb.bio.ic.ac.uk/ecotron/

Speciation in action

Byrne, K. and Nichols, R.A. (1999) *Culex pipiens* in London Underground tunnels: differentiation between surface and subterranean populations. *Heredity*, 82: 7–15.

Hoch, H. and Howarth, F.G. (1999) Multiple cave invasions by species of the planthopper genus *Oliarus* in Hawaii (Homoptera: Fulgoroidea: Cixiidae). *Zoological Journal of the Linnean Society*, 127: 453–475.

Korol, A., Rahkovetsky, E., Iliadi, K., Michalak, P., Ronin, Y. and Nevo, E. (2000) Nonrandom mating in *Drosophila melanogaster* laboratory populations derived from closely adjacent ecologically contrasting slopes at 'Evolution Canyon'. *Evolution*, 97: 12,637–12,642.

3 An inventory of planet Earth

Discovering and defining the richness of life is central to our knowledge of biodiversity. This chapter covers:

- **Defining types of biodiversity**
- **Quantifying biodiversity**
- **Global patterns of biodiversity**
- **The role of ecosystems using the example of wetlands**

Biodiversity strikes a Pavlovian nerve, conjuring up images of rare mammals and rainforests but it embraces much more than these. There are three main components to biodiversity: genetic, organismal and ecological. The **genetic and subcellular diversity** includes all biodiversity expressed within individual cells plus non-cellular organisms such as viruses. The diversity of genetic information is central to this category but the variety of metabolic pathways and molecular biology of life also represent important diversity. **Taxonomic diversity** is dominated by our focus on species. Species are but one level at which organisms can be classified and the diversity of other categories such as families or phyla provides additional insights. **Ecological diversity** includes whole communities, habitats and ecosystems, including domestic stocks. Various definitions including UNEP's Global Biodiversity Assessment have added **cultural biodiversity** as an explicit concept, human social systems intimately dependent on the ecological system within which they exist. This has created some tensions. In 1996 members of the Makah nation attended the International Whaling Committee meeting to request permission to hunt five grey whales a year, after a seventy-year gap, to help maintain their cultural system. Their request was seized upon by Norway, keen to reopen commercial whaling. Small Norwegian coastal towns that would benefit from commercial whaling were compared to the Makah people. **Interaction biodiversity** has also been coined to embrace the interactions between species as a fundamentally important factor in natural systems. There are at least eighty-five published definitions of the word and the very success of the term biodiversity, its rise to prominence and increasing breadth of topics have fuelled some dissent, with scientists accused of pushing the term as a technological, even mythic concept, because of its allure for research funding (see Table 3.1).

In 2003 a review of the World Wide Web hit upon 3.1 million sites with the word, more than other major environmental or biological fields such as molecular biology. This usage was even described as 'market penetration'. Other authors are less flattering. Peter Marren, a veteran of UK conservation, introduces the concept as 'a characteristic modern compound-noun combining "biological" and " 'diversity" . . . when traditional English was still spoken we would have referred to it as "the variety of life on Earth".'

Part of the problem is the glee with which bureaucracies have seized on the new jargon. Additionally the word embraces so much as to sometimes appear useless: 'Biologists are

Table 3.1 Categories and definitions of biodiversity

Author	Genetic and subcellular	Taxonomic	Ecological
Eldredge 1992	Genealogical	Phenotypic	Ecological
Groombridge 1992	Genetic diversity	Species diversity	Ecosystem
Department of the Environment 1994	Genetic variation	Diversity of species	Diversity between and within ecosystems
Harper and Hawksworth 1995	Genetic	Organismal	Ecological
Heywood 1995	Genetic	Organismal	Ecological

inclined to agree that it is, in one sense, everything', as E.O. Wilson wrote in the Introduction to *Biodiversity II*, a sequel to the 1988 classic (Reaka-Kudla *et al.* 1997: 1). He followed up with this more precise definition:

> Biodiversity is defined as all hereditarily based variation across all levels of organization, from the genes within a single local population or species, to the species composing all or part of a local community, and finally to the communities themselves that compose the living parts of the multifarious ecosystems of the world.
>
> (Reaka-Kudla *et al.* 1997: 1)

However, the complexities of technical definitions are often bettered by the understanding of the term in a wider social context. Here are two of my favourites:

> Biodiversity is our life insurance.
> (Peter Schei, Senior adviser to the UN and Directorate for Nature Management, Norway)

> Biodiversity is the safety net for many of the world's poor.
> (Achim Steiner, IUCN Director General)

Types of biodiversity

Genetic and subcellular

Genetic, molecular and metabolic diversity are often overlooked. They represent the founding diversity of life, not just as it is nowadays but with echoes of the ancient past and potential futures. All three generate diversity within and between species and ecosystems.

Genetic diversity provides the core differences that divide life into its major types, especially important when visible structures and shape provide no reliable guide. This genetic diversity is witness not only to the deepest divisions of life but also to some shared characteristics, stretching across huge taxonomic distances. Such genetic characteristics speak of the relatedness of all life forms one to another. Genetic diversity is the ultimate divisor and link. Genetic and metabolic diversity are especially important to understanding time, evolution and taxonomic characteristics of life. Genetic material is the raw material from which future biodiversity will be spawned.

Genetic diversity is made up of **genes**. A gene controls the expression and development of a particular feature of a living organism. The precise form of the character will vary (e.g. hair or eye colour in humans) and each of the variations of the gene is called an **allele**. So a gene in an organism will be one allele from a larger set. Genes are built of **deoxyribonucleic acid (DNA)**, a linear molecule built like a ladder and twisted into the famous double helix. The rungs of the ladder consist of molecules called **nucleotide bases**, two joined to make each rung. There are four different bases, cytosine, guanine, adenine and thymine, which pair up to build the rungs as base pairs. Cytosine pairs with guanine, adenine with thymine. Each set of three base pairs makes up a triplet, and each triplet is the code sequence for an amino acid. There are sixty-four permutations in which three consecutive base pairs can occur, more than enough to code for the twenty amino acids common to all organisms. So the diversity of the genetic code is a hierarchy starting with the sequence of individual rungs then triplets, the order and length of sets of triplets, the resulting alleles and quantity of genetic material in different organisms. In addition to the active genes there is extra DNA apparently doing nothing.

The result is that genetic diversity even within a single species is so vast that the information is much larger than the total number of individuals. The measurement of genetic diversity is a rapidly improving area. Box 7 outlines methods and measures.

Box 7

Genetic biodiversity: methods and measures

Methods

- **Protein electrophoresis** Widely used since the 1960s, this technique analyses the different proteins which in turn reflect the different alleles in an individual.
- **Restriction site mapping** This recent advance relies on very specific bacterial enzymes that restrict viral damage to DNA by cutting damage out at specific points. Analysis of these cut points, called restriction sites, allows very precise analysis of gene sequences.
- **DNA and RNA sequencing** Another new approach that allows analysis of all DNA. A frequent alternative is ribonucleic acid (RNA), found in ribosomes (rRNA). Ribosomes are cell organelles that read genetic data and manufacture proteins based on these instructions; 16s rRNA (the name refers to the number of subunits in the RNA and that it is ribosomal) has been particularly important.

Measures

- **Percent polymorphic loci, P** A locus is the position of a gene on the genome. A gene where the frequency of the most common allele is <95 per cent of the total is regarded as polymorphic.

- **Number of alleles, N**   This measures not only if a gene is polymorphic but also the number of alleles per gene.
- **Heterozygosity, H**   Frequency of alleles, how many there are and the frequency of each, e.g. there may be three alleles, one occurring 85 per cent of the time, a second 10 per cent, a third 5 per cent.
- **Number of segregating sites, S**   Restriction site positions can vary in a gene. This detail can be added to the first three items.
- **Allele tree**   This is an attempt to construct evolutionary relationships, providing a sense of relatedness and uniqueness. In much the same way as a rare species can be picked out from amongst its more common kin, allele trees try to pick out the genetically unusual.

The metabolic diversity of life, particularly amongst bacteria, is commonly overlooked, but represents a fundamental and ancient diversification (see Box 8). The metabolic abilities of bacteria can be directly useful to humans (e.g. fermentation) or indirectly as part of wider geochemical cycles that are ecosystems services maintaining the health of the environment. Metabolic diversity can be divided between utilisation of different **energy sources**, **energy release** (respiration) and **nitrogen fixation**.

Box 8

Microbial metabolic diversity

Tapping environmental energy sources.

Energy from inorganic chemistry (chemoautotrophy)

- *Sulphur bacteria:* oxidise sulphur compounds, e.g. hydrogen sulphide, to release energy used to build carbon-based food.
- *Iron bacteria:* oxidise iron compounds, e.g. iron ore, to release energy used to build carbon-based food.
- *Nitrifying bacteria:* oxidise ammonia to release energy used to build carbon-based food. Nitrite and nitrate by-products are a valuable input of nitrogenous plant nutrients.
- *Hydrogen bacteria:* oxidise hydrogen to release energy used to build carbon-based food.
- *Athiorhodacea:* split organic matter as hydrogen ion donor.
- *Purple and green sulphur bacteria:* split hydrogen sulphide as hydrogen ion donor. Waste product is sulphur dioxide.
- *Advanced photosynthesis:* split water as hydrogen ion donor. Waste product is oxygen.

Note that many bacteria can trap energy by photosynthesis and by oxidation of inorganic molecules. The ability to use either method is called mixotrophy.

Metabolic energy release

Anaerobic

- *Fermentation:* partial breakdown of carbon compounds.
- *Anaerobic photosynthesis:* organic matter broken down as hydrogen donor for photosynthesis generates energy.
- *Nitrate reducers*: split oxygen off nitrate and nitrate. Nitrogen gas produced.
- *Carbonate reducers:* split oxygen off common carbonate compounds, e.g. calcium carbonate. Waste product, methane, has a possible role as a greenhouse gas.
- *Sulphate reducers:* split oxygen from sulphate. Create 'rotten egg' hydrogen sulphide smell of fetid wetlands.
- *Iron reducers:* alter chemical bonds of iron compounds resulting in energy release analogous to oxygen splitting of other anaerobic systems but without oxygen.

Aerobic

Oxidation of organic matter: primary energy release systems for protistan, fungal, plant and animal kingdoms.

Nitrogen fixation

Bacteria extract atmospheric nitrogen gas and make it available to nitrogen cycle. Major source of nitrogen input into ecosystems.

Molecular biodiversity is less familiar but offers an important alternative approach to describing sub-cellular variety, which is important for food, pharmaceuticals, cosmetics and industrial compounds (e.g. dyes). Molecular biodiversity can be described using molecular structures or functions (e.g. growth stimulant, anti-microbial, alarm) and audited by habitat, biogeographically region or species. For example, South American amphibians are characterised by their variety of steroids, temperate grasslands their morphine derivatives and northern conifer forests their terpenes. Although not such a familiar image of biodiversity as pandas and rainforest, the new analytical and informatics techniques are rapidly allowing molecular biodiversity to catch up as a significant approach to understanding the richness of life on Earth.

Taxonomic diversity

The number of species currently alive on Earth is often thought of as synonymous with the term biodiversity. Work at the species level has practical advantages. Funding bodies (often the general public) recognise species. Conserving a species is a simpler concept than conserving a specie's gene pool. Species projects are often easier to organise than conserving whole habitats. However, the species is but one category in a hierarchy of classification. Other categories are important measures of biodiversity sometimes providing different, contradictory messages to species-level analyses.

Systematics is the branch of biological science responsible for recognising, describing, classifying and naming organisms. Part of this work involves a **taxonomic** (taxa = category, nomic = name) hierarchy, including species and other levels at which organisms can be described. Linked to this are taxonomic nomenclatures (naming systems) which provide internationally agreed rules on how species are to be named. International codes exist for animals, plants and bacteria, plus there is a provisional system for viruses. The taxonomic classifications also portray our understanding of the evolution and relatedness of life, the **phylogenetic classification**. Humans have a natural tendency to classify things, biodiversity being no exception. Classification depends upon what we can observe and what we think it means (often influenced by social, political and religious attitudes). Taxonomic classifications of life have varied throughout time. Box 9 outlines the development of schemes in Europe.

Box 9

Classifications of life through European history

Classifications of life have changed over time, but even ancient systems can show precise methodologies and criteria. Changes reflect use of increasingly small scale, internal features (internal organs, internal cell organelles, internal organelle RNA).

Greek

Aristotle, in *Historia Animalia* (486 BC), classified animals by clearly stated criteria for grouping animals together: (1) 'With regard to animals there are those which have all their parts identical . . . specifically identical in form'; (2) 'When other parts are the same but differ from one another by more or less, then they belong to animals of the same genos. By genos I mean for example bird or fish'; (3) 'There also exist animals whose parts are neither the same by form but by analogy'.

Linnaeus

The famed Swedish biologist tried a global classification of animals using multiple criteria. Life forms must be distinguished by external criteria, e.g. skin, locomotion. Each life form should have roughly equal roles in the economy of nature. Within each life form criteria used for subgroups should be essential for finding or processing food, e.g. teeth for mammals, beak for birds. Linneaus' work resulted in classifications of plants based primarily on reproduction mechanisms and of animals based on feeding.

Cuvier

Through the late eighteenth and early nineteenth centuries Cuvier pioneered comparative anatomy, especially the internal resemblance of animals, as a means of classification. Cuvier saw some characteristics as fundamentally more definitive than others, notably the nervous systems, though practicalities favoured use of bones. Cuvier broke the vision of animal life as a ladder of progression, splitting animals into four lineages – vertebrates, molluscs, articulates (e.g. insects) and radiates – based on form and function. This split is still echoed in current classification of animal lineages.

Whittaker's five kingdoms

By the mid-twentieth century advances in cell biology led to a revision of life into five main kingdoms defined primarily by subcellular features, e.g. cell walls, links between cells, organelles and mechanisms of cell division. The five kingdom system (Monera, Protista, Fungi, Plantae, Animalia) is still widely used (see Figure 3.1).

16R rRNA

By 1990 analyses of DNA and RNA, particularly work on 16s rRNA, sequences suggested a fundamental revision of the five kingdoms. Shared and different 16s rRNA sequences suggested that life divided into two main branches, the Bacteria on one hand and a second branch itself split between the Archaea (previously thought of as close allies of the Bacteria) and the Eucarya (everything else). These three groups are now called domains, of fifteen, three and twenty-one kingdoms respectively. Animals, plants and fungi are but three of the kingdoms within the Eucarya. Recent evidence suggests that the relationships between the three domains are muddied by transfer of genes between ancient lineages that eventually became isolated as distinct kingdoms. This 'New Tree of Life' suggests that the ancestors of all life on earth may be more a community of promiscuous microbes than a single lineage.

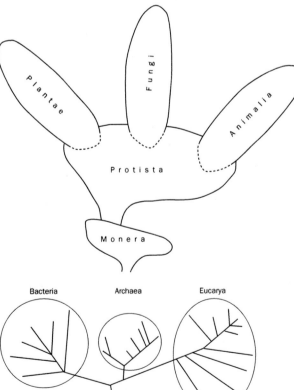

Figure 3.1 Whittaker's five kingdoms of life and recent three domain revision based on RNA. The Bacteria and Archaea domains were both part of Whittaker's Monera, while the new Eucarya domain encompasses what were the Protista, Fungi, Animalia and Plantae Kingdoms

Source: Redrawn from Heywood (1995).

Table 3.2 The taxonomic hierarchy. Each level in the hierarchy is a category at which fundamental features are shared. Each category will consist of one or more of the next category down (e.g. a genus consists of one or more species). Sometimes subcategories are used (e.g. subclass, suborder). Recent analysis of genetic biodiversity has caused a rethink of what had seemed the ultimate, kingdom-level divides, a salutary lesson

Taxonomic hierarchy	Human beings	Mount Kupe bush shrike (Chapter 4)	African elephant (Chapter 4)
Kingdom or domain	Kingdom Animalia; Domain Eucarya	Kingdom Animalia; Domain Eucarya	Kingdom Animalia; Domain Eucarya
Phylum or division	Chordata	Chordata	Chordata
Class	Mammalia	Aves	Mammalia
Order	Primates	Passeriformes	Probosidea
Family	Hominidae	Corvidae	Elephantidae
Genus	*Homo*	*Telophorus*	*Loxodonta*
Species	*sapiens*	*Telophorus kupeensis*	*africana*

Recently systematics and **taxonomy** have not been popular fields with biologists or funding bodies but burgeoning biodiversity studies have focused attention on our woefully incomplete catalogue of life. 'You don't know what you've got 'til it's gone', the 1960s environmental protest song lamented. We now realise that you don't even know then. An urgent task in biodiversity research is to rebuild skills in systematics. Table 3.2 gives the typical classification hierarchy, from species up to the fundamental divisions of life into kingdoms or domains. Note that there is marked variation in diversity even at very high levels of classification. Recognised numbers of phyla in the five kingdom classification of life are: Prokaryte 16, Protista 27 (at least), Fungi 5, Animalia 32 and Plantae 9.

Defining species

The **species** remains the focus of most biodiversity research, but there is no precise, simple, universal definition of what a species is. Some alleged species fit all criteria of all definitions, others are hard to fit to any.

The word species literally means outward or visible form. Classifications up to the twentieth century relied on the physical, often outward, similarity of features to distinguish a species. The use of visible characters to define a species is the **morphological species concept**. This approach is widely used with collections of dead specimens or fossils which can provide no other information on breeding and ecology. Species with varied individuals, e.g. camouflaged, mimics or variable growth forms, risk being split into several species.

Insights from evolution and genetics prompted the **biological** (or **isolation**) **species concept**. A species consist of populations of interbreeding individuals, able to reproduce successfully with other populations. Central to this definition is the idea of reproductive isolation, be it a physical barrier or behavioural, physiological or genetic inability to mate and produce fertile young. The isolating mechanisms can be **pre-** or **post-mating**. Pre-mating includes separate habitat, time, behavioural and physiological mismatch, mechanical inability to mate and gamete inability to fertilise. Post-mating includes hybrid inviability (embryos or young die), hybrid sterility (young are born but cannot reproduce) and hybrid breakdown (young will mature and breed but their offspring are less viable). This definition does not work with asexual species. A surprising number of recognised species living side by side appear to mate and exchange genetic material without the species involved coalescing into an indistinguishable melange. Many plants show interbreeding. Tests of the American buffalo genome have revealed domestic cattle genes but no morphological blur. Species defined by breeding are also difficult for systematists. Not only are the breeding abilities of fossils impossible to fathom but also over evolutionary time species change, so much so that they might be morphologically distinct, but at every step interbreeding between immediate parent–offspring generations was possible. The morphologically distinct ancestor and descendent species are linked by a seamless reproductive thread.

Refinements to the biological species concept include the following: the **recognition species concept**, emphasising not the barriers to reproduction but shared characteristics permitting reproduction, the **cohesion concept**, a species defined by cohesive mechanisms that prevent interbreeding with other species and evolutionary diversification within the species, and the **ecological concept**, species defined by occupation of a niche unique compared to other species in its range and evolving separately from populations beyond its range. All remain flawed by hybridisation and problems of practical measurement.

A third approach is the **evolutionary species concept**. This emphasises the **phylogenetic concept** with species defined as the smallest population (sexual) or lineage (asexual) diagnostically distinct from other such populations and with a discrete lineage. The diagnostic characters are often genetic. When applied to species recognised by morphological or biological concept criteria, the sheer genetic diversity within species often suggests they should be splintered into many separate lineages. The evolutionary concept defines a species by a single, distinct lineage with an identifiable evolutionary history and fate.

Ecological diversity

Taxonomic diversity classifies types of organisms and their relatedness but organisms do not live in isolation from one another or the physical world. Humans have long recognised different ecosystems of apparently interdependent life. Ecological science focuses on these patterns and processes, hence ecological diversity is the inclusive term for this third category.

Ecological diversity covers a host of concepts; ecosystems, communities, assemblages, habitats, biomes and biogeographical regions. These are not one and the same, indeed some may not be biodiversity proper. The term **ecosystem** embraces the living organisms and non-living (abiotic) features such as climate and geology of a site. Some have argued that the inclusion of abiotic components excludes ecosystems from biodiversity. A **community** refers to organisms living together, essentially the live component of an ecosystem. Again this deceptively simple idea is problematic. The term implies a linkage, an interdependence of species that may not exist. Even tightly linked communities will harbour fleeting tourist species, moving through without necessarily playing any role whilst other communities may be very loosely tied assortments, often described as an **assemblage**. Habitat conjures up precise images such as the giant panda habitat, but the term may not mean anything if the species is not present; does giant panda habitat cease to exist should the giant panda become extinct? **Biome** is the term associated with global or continental scale, regional ecosystems defined by vegetation and fauna, in its turn largely determined by climate. A **formation** is a similar concept, relying solely on vegetation data. Even colloquially familiar biome terms such as rainforest or wetland become difficult to define precisely due to global variations or to give sharp boundaries where they grade into each other.

Ecological biodiversity also includes geographical foci such as high diversity hot spots, continental and oceanic islands plus regions of endemism. If ecological diversity (whether

ecosystem, community or any other tricky concept) was nothing more than the s
parts (e.g. the species list) this category of biodiversity could be largely ign
ecological diversity has its own importance. Ecological patterns emerge at the e---,
and community level which are more than the sum of the parts of the species present.
Ecosystems can (albeit rather mechanically) be said to provide ecosystem services
(or functions) such as nutrient and gas cycling. An ecosystem that loses a species may
not be merely minus one species but may function in a very different way with knock-on
consequences to other ecosystems and perhaps globally. So the ecological category of
diversity is important.

Classifications of ecological diversity vary with scale, just like the genetic (nucleotide
molecules up to differences between populations) and taxonomic (subspecies to domains).
Three main scales are commonly for used ecosystems: global, regional and national. Global
classifications rely on major vegetation types (biomes) typically defined by combinations
of dominant vegetation, landscape and climate, sometimes plus biogeographical position.
Regional classifications are often based on practical definitions. National classifications
work at a finer scale, often based on precise species presence and abundance. Classifications
of wetlands (global, regional or national) are good examples of ecosystem categorisation,
the difficulties of definition and how resulting classifications can vary with the purpose
of the systems used.

Wetland classification

Global classifications

At the gross, global level, wetland habitats are subsumed within the broader biomes. A
widely cited definition of wetlands is from the Ramsar Convention (the treaty is outlined
in Chapter 5):

> Areas of marsh, fen, peatland or water, whether natural or artificial, permanent
> or temporary with water that is static, flowing, fresh, brackish or salt, including
> areas of marine water the depth of which at low tide does not exceed six metres.

Wetlands are characterised by the presence of water at the surface or in the root zone,
unique soil conditions and hydrophyte (water loving) plants. Distinguishing different types
of wetlands is problematic. The duration and depth of inundation vary. Wetlands are often
marginal habitats between truly aquatic and terrestrial systems. Size can vary from small
ponds to huge regional expanses. Definitions also vary according to the aim of the
classification scheme and cultural traditions, e.g. in the United States, swamp means
forested, in Europe typically not so. At a global scale nine main wetland types are commonly
distinguished using general biome and landscape characters.

- *Bogs:* acidic, peat-building wetlands with water and nutrient input from precipitation,
  typically in wet, cool climates and characterised by *Sphagnum* mosses.
- *Fens:* peat-building but with nutrients and hydrology influenced by surrounding
  landscape and geology, neutralising acidity. (Note that bogs and fens are often
  combined together as **mires**, a generic term for peat-based wetlands.)
- *Swamps:* so waterlogged that water usually covers surface; dominated by a few species
  e.g. reeds or characteristic trees, e.g. swamp cypress.
- *Marsh:* inundation often seasonal: diverse, herbaceous vegetation with little or no peat
  accumulation.
- *Floodplains:* periodic over-spill from lakes and rivers; very varied.
- *Shallow lakes:* open water up to a few metres deep.

- *Salt marsh:* tidal herbaceous coastal sward of temperate latitudes.
- *Mangroves:* tidal coastal woodlands between 32°N and 30°S, characterised by mangrove trees; exclude saltmarsh from tropics by competition but limited to tropics by climate.
- *Anthropogenic:* wetlands created by humans, e.g. paddies.

National classifications of wetlands

At a national level much more detailed classifications can be devised, often differing depending on the purpose for which they are used. In the United States there are over fifty wetland classifications systems in use for water regulation, recreation and wildlife conservation. Here is an example from the UK based on detailed botanical classification.

Classification by species and floristic composition: the National Vegetation Classification

The National Vegetation Classification (NVC) is a systematic and comprehensive classification of UK plant communities. Vegetation types have been distinguished by computer analysis of the presence and absence of individual species and their relative abundance. The classifications for four habitat types (woodland, mires, grassland and aquatic) are available. All four contain communities identified as wetland, a hundred in all, with coastal ecosystems still to be added. Each community type is described in detail, listing plant species, abundance, physical and chemical conditions in the habitat and distribution. Each type is given its own code and name, derived from the diagnotistic vegetation. For example, S9 *Carex rostrata* swamp is the ninth recognised swamp community from the aquatic habitats, characterised by the bottle sedge, *Carex rostrata*.

- *Woodland:* seven wet woodland communities
- *Mires* (bogs, wet heaths and fens): thirty-eight wetland communities
- *Grassland:* three inundation communities
- *Aquatic:* twenty-four submerged and twenty-eight swamp/tall herb fen communities.

Regional classification of wetlands

The SADCC Wetlands Programme is a wetlands inventory for southern Africa. In 1991, the Southern African Development Co-ordination Conference (SADCC) Wetlands Conservation Conference, meeting in Gabarone, capitol of Botswana, set up a programme intended to produce a comprehensive inventory for the ten countries of Angola, Botswana, Lesotho, Malawi, Mozambique, Namibia, South Africa, Swaziland, Zambia and Zimbabwe. The driving force was increasing recognition of the value of wetland ecosystem biodiversity: wetlands have the highest biological productivity compared to equal areas of other ecosystems in the region, they offered multiple uses and harboured the greatest taxonomic biodiversity. The programme collects baseline data on wetland area, distribution seasonality, characteristics and value at a national level. The national scale is important to draw up reasonably clear boundaries and to detect small and unusual combination wetlands. Sources of data included large-scale satellite remote sensing, aerial photographs, maps and vegetation surveys, as well as collating existing reports on individual wetlands. Highly detailed botanical surveys are lacking or insufficient to create classifications as detailed as the UK National Vegetation Classification but this is not a serious problem given the purpose of the SADCC inventory. The inventory work also produced practical benefits, improving co-operation across political borders that cut through wetlands and between different interest groups such as farmers, foresters and conservationists. The final aim is threefold, to quantify the extent, types and uses of wetlands, to quantify the rate and extent of alteration and loss and thirdly to disseminate the information to managers and users of these precious ecosystems. Wetland classifications in Zimbabwe reflect the different ecosystems and human concerns.

Zimbabwe's wetlands occupy about 3.3 per cent of the country. No systematic classification exists as yet but the broad categories are swamps, permanently inundated; floodplains and flats, seasonally inundated; dams, manmade, small impoundments important for irrigation; pans, seasonal waterholes sometimes supplied by pumped water for cattle or wildlife tourism; **dambos**, small wetlands widely used for horticulture, and lakes. Zimbabwe is landlocked and has no coastal ecosystems.

Domestic biodiversity

Biodiversity is widely thought of as an entirely natural phenomenon but definitions now recognise the importance of domesticated species, ecosystems forged from human activity and some cultures intimately tied to their environment. Biodiversity plays a role in the ethical, religious and social values of societies. Differences between cultures affect our valuation of biodiversity, in turn affecting conservation. The fate of biodiversity in different cultures may provide important insights since humans are now the dominant influence on remaining species and ecosystems. Several indigenous cultures are (or were) founded on an ecological intimacy with biodiversity. The native plains tribes of North America relied on the buffalo herds, not only as food but also as an icon central to their beliefs and customs. The Marsh Arabs of Iraq have a social and agricultural system dependent on their wetland home.

Domestic biodiversity is a tiny fraction of the whole but provides 90 per cent of human food supplies. Domestic stocks include life from all kingdoms. In addition landscapes created by humans, whether farmland, forests, gardens and urban landscapes, are new ecosystems. The extent of human interference with domestic biodiversity varies. Truly **domestic species** have been subject to marked **artificial selection**, so that many characteristics will be markedly different from wild ancestors such as cattle, sheep and pigs. **Domesticated species (exploited captives)** retain many similarities to wild brethren, e.g. reindeer, yak, camels. **Feral** populations are domesticates that have reverted to living wild. They can create problems, for example the feral cat predation of native marsupials in Australia, but are unique additions to wildlife. Domestication leads to a proliferation of forms, **cultivars** in plants and **breeds** in animals, which can be defined by heritable, distinct and uniform characteristics different from other such groups. Plant domestication spans five categories, first the **wild ancestor** stock. **Weedy relatives** often flourish in marginal habitats impacted by humans, a genetic bridge between wild relatives and the truly domestic. **Primitive cultivars** are genetically diverse stock still open to natural selection. **Modern cultivars** are the result of purely artificial selection of supermarket qualities. **Advanced breeding lines** are the recent additions, their genomes created by laboratory manipulation, perhaps including genetic diversity impossible through natural processes.

Numbers of domesticated species are small. Of the 320,000 vascular plants, 500 are domesticated and 15–20 are major crop species. The 50,000 vertebrates (excluding fish) have provided 30–40 domesticates, plus 60 semi-domesticated. Another 200 fish and invertebrates are commonly used in aquaculture and 4 insects (honey bee, various silkworm caterpillars and the cochineal bug). Fungi and microbes are grown commercially for food, brewing and biotechnological uses (e.g. pharmaceuticals). The lack of species is made up for by breeds. Estimates of total numbers of breeds vary, for example cattle 780–1090, sheep 860–1200, pigs 260–480 (if we cannot keep track of domestic diversity is it any wonder that numbers of wild species are so difficult to estimate). The core domestic farm mammals (cow, buffalo, horse, ass, pig and sheep) comprise 3237 breeds of which 474 are rare and an additional 617 have gone extinct since the end of the nineteenth century. The variety of breeds is a resource. Many have adaptations to local conditions, these adaptations can be exported to new ranges either by moving the animals or controlled breeding. The known traits of breeds allow easier control of artificial selection and discovery of new mutations. The variety is also insurance against the unforeseen. However, this richness has often been overlooked, the potential value of domestic variety suddenly revived in the face of major threats such as climate change. The UK government agencies have caught up with the problem, releasing a report in 2003 examining the genetic resources for food and agriculture which emphasised the need to conserve and develop Britain's traditional breeds, including securing breeds at risk in gene banks and examining the wealth of adaptations to local conditions which they represent. The variety is being eroded by the globalisation and intensification of agriculture. Losses are biased towards the developed world which may have nurtured more breeds and recorded extinctions. Table 3.3 gives global, regional (Europe and Africa) and national (UK and Zimbabwe) cattle and sheep breed diversity and losses.

The inventory of domestic diversity also includes more ominous items. Manipulations of genes in some animals and plants have created hybrids which would probably be impossible through natural selection. Incorporation of genes from one species into another is now commonplace, creating **transgenic species**. Transgenic sheep, pigs, cattle, horses and rodents have been created containing genes to enhance obvious features, for example mouse genes controlling fur structure in sheep enhance fleece productivity. Human genes have also been incorporated into some animal species to produce medical products such as growth hormones. Genes can even be combined between kingdoms with bacterial genes incorporated into some mammals. Such advances now include the ability to coalesce species, most famously the goat/sheep hybrid which is an example of a **chimera species**.

Disease eradication programmes have also created a dilemma for conservation. One virus, smallpox, once a deadly killer, was declared extinct in the wild by 1980. Two stocks remain in laboratories, one in the United States and one in Russia. In 1995 the World Health Organisation (WHO) voted not to destroy these stocks. Had the destruction been authorised, this would have been the first ever intentional extinction of a species. In 1996 the WHO recommended destruction in June 1999 following a three-year search to check for forgotten or hidden stocks, nervous of illegal stockpiles kept for biological warfare. Instead the intrinsic value of the virus, as well as possible future need for research in the event of the rediscovery of a wild stock or near relative, brought a reprieve.

Table 3.3 Surviving and extinct cattle and sheep breeds. Numbers given for global, European, African, UK and Zimbabwean records

	Global	Europe	Africa	UK	Zimbabwe
Cattle					
Extant total	732	209	168	42	4
Extant rare	141	154	10	16	0
Lost	224	101	22	5	3
Sheep					
Extant total	917	356	131	73	3
Extant rare	135	109	1	17	0
Lost	146	97	4	8	0

Sources: Groombridge 1992; Heywood 1995

Chimeras and deadly viruses are biodiversity's stranger corner but our attitudes to nature have always been coloured by the culture of the day. One Anglo-Saxon bishop issued a letter to his parishes scolding the populace over their fear of werewolves which were, after all, a common feature of the countryside. Given the welter of definitions, aspects and scales of biodiversity the key is always to be precise about exactly which component you are exploring. Box 10 uses the example of UK temperate ponds to illustrate how different aspects of biodiversity can be explored within one type of ecosystem.

Box 10

Types of biodiversity: the example of ponds in the United Kingdom

Ponds are a common habitat throughout the United Kingdom, indeed in most countries. The biodiversity of ponds can be investigated from different perspectives.

Genetic biodiversity

Ponds are scattered across the landscape and the wildlife of ponds disperses with varying degrees of effectiveness between sites. Many of the smaller crustacea of ponds such as water fleas (Cladocera) and pea shrimps (Ostracoda) reproduce parthenogenticially so that populations develop as clones from the initial colonists. Therefore individual ponds may support distinctive clones, perhaps based on a limited genetic bottleneck of only a handful of original colonists.

Taxonomic biodiversity

Ponds are small habitats compared to lakes and rivers. National surveys have revealed that British ponds are disproportionately rich in species of plants macroinvertebrates, supporting more species than found in lakes and river systems.

Ecosystem biodiversity

A common, if rather befuddling, characteristic of temperate ponds is the variation of animal and plants communities between ponds that are very close together and apparently developing in the same climatic and geochemical environment. Why this heterogeneity should develop is not clear. It may be down to historical accidents of which species arrive first and take over or it may reflect much more finely tuned responses to small variations in the environment.

Domestic biodiversity

Many ponds have been created by people. There are numerous reasons, e.g. drinking water, fire fighting, defence or decoration. Different types of ponds can create different opportunities for wildlife. Decoy ponds to lure in wildfowl for food may be excellent for other wildlife such as attracting ducks. Garden ponds have become not only an important refuge for amphibians but also a source of invasive species of non-native plants.

Cultural biodiversity

Ponds have an ancient place in British culture, with similar echoes throughout the world. Ponds and similar sites such as springs are often seen as portals between our material world and spirit worlds. In the United Kingdom this is evident as far back as the Bronze Age (AD 3000) from finds of rituals offerings, often immaculate and costly weapons, placed into ponds. This tradition survives in our custom of throwing coins into wishing wells.

Measuring biodiversity

Quantifying the variety and richness of life on Earth is central to studies of biodiversity. This section provides estimates of taxonomic and ecosystem diversity but be warned, the numerical data are our best guesses, not definitive answers. For example, Table 3.3 of surviving cattle and sheep breeds disguises serious differences between estimates. If we cannot be sure about numbers of animals which we control, estimating numbers of wild species is even more difficult.

The numbers of known species

Inventories of the present-day diversity of species on Earth can use known species and estimates of likely numbers (including likely maxima and minima). The numbers known and reliability of estimates vary greatly with taxa. Birds and mammals are well documented, though there are still some surprises, e.g. the Udzungwa partridge (*Xenoperdix udzungwensis*), a bird species first identified in 1991, in a dinner (see Box 11).

Numbers of many invertebrates and micro-organisms remain darkly mysterious. Estimates of the rate at which new species are discovered can provide some insight into these voids. Table 3.4 gives estimates of known species, showing how these have changed through time.

Estimates of global diversity have changed markedly throughout history. It is important to separate these counts of known taxa from estimates of the actual numbers, which use various assumptions to include projections of likely numbers as well as the as yet undescribed.

The total described inventory for some taxa such as multicellular animals and green plants gives good minimum estimates, but there are problems even with these organisms we have described. For some there is a debate over the numbers of genuine species versus subspecies. In more diverse, typically less well-known taxa, including some of the numerically and ecologically dominant groups such as worms and molluscs (Phyla Annelida and Mollusca respectively), there is confusion because several nominal species (anything that has been described and had a scientific name attached to it) are actually specimens of the same species. Each of the names is merely a synonym for the same thing, perhaps arising because of variable shapes, names given to different stages in the life cycle, earlier names remaining unknown to later taxonomists or confusion due to reclassification. The numbers of known mollusc species varies between 45,000 and 150,000, with 70,000 a compromise. Extensive, tangled synonyms need painstaking revision. Even without these problems, the literature of described species is confused and scattered. The apparently simple task of totting up published records gives different totals in different studies (see Figure 3.3).

Box 11

A new species of bird: the Udzungwa Partridge

Our knowledge of bird species is unusually complete but new species can be found. In 1991 two Danish zoologists were exploring Udzungwa Mountain, part of the endemic rich Eastern Arc Mountains of Tanzania. One night at supper the expedition cooks produced 'chicken stew'. At the bottom of the pot were the two legs of a bird, not exactly a chicken (African chicken stew is often not exactly chicken) but which the zoologists thought must be one of the chicken-like Francolins. Keeping an eye out for Francolins, not previously known in the area, subsequent sightings of the bird did not match anything in the published literature. During a return visit in search of the mystery bird, local guides snared a male and female. Comparison of the specimens to museum skin collections revealed that the bird was a new species, related to species from South East Asia. The Udzungwa forest partridge (*Xenoperdix udzungwenis*) was not merely a new species of bird, but a relict species from 15 million years ago when forests spanned Africa to Asia (see Figure 3.2). The Udzungwa partridge is a classic relict species, contributing to the diversity of this hot spot. In 2001 a second population was discovered on mountains 150 km north of the Udzungwa sites; these new birds were perhaps different enough to count as a new species. By 2004 the partridge was reclassified from Vulnerable to Endangered by the IUCN, with an estimated population of 3700 scattered over four mountains, although more recent surveys failed to find it at two of these sites. Snaring and disturbance may be significantly reducing the local remnant populations.

Figure 3.2 The Udzungwa Partridge

Source: Reproduced with kind permission from Martin Woodcock and the African Bird Club.

Table 3.4 Numbers of described species and estimates of actual numbers for selected taxa (in thousands)

Taxa	Species described	Estimated numbers of species: high	Estimated numbers of species: low	Working figure
Viruses	4	1,000	50	400
Bacteria	4	3,000	50	1,000
Fungi	72	2,700	200	1,500
Protozoa and algae	80	1,200	210	600
Plants	270	500	300	320
Nematodes	15	1,000	100	400
Insects	950	100,000	2,000	8,000
Molluscs	70	200	100	200
Chordates	45	55	50	500

Sources: Groombridge 1992; Heywood 1995

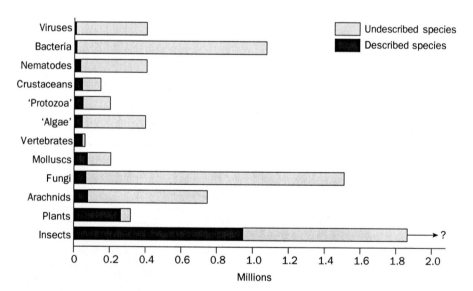

Figure 3.3 Estimates of described and likely totals of species for selected taxa

Source: Redrawn from Heywood (1995).

Species are still being described, even for the best known groups, and description rates provide some sense of the task ahead. Analysis of standard published records of new species between 1979 and 1988 shows remarkable consistency with around 10,912–11,599 new species of animal added each year. As a general figure 13,000 new species are described per year. Completely new phyla may turn up. In 1993 an extraordinary animal was found living on the mouthparts of the Norway lobster (*Nephrops norvegicus*). The individuals attached to the lobster are one of several attached or mobile stages in a life cycle including sexual and asexual reproduction. It appears to feed by filtering scraps of food discarded as the lobster chews. The creature has been named *Symbion pandora*, and the new Phylum named the Cycliophora.

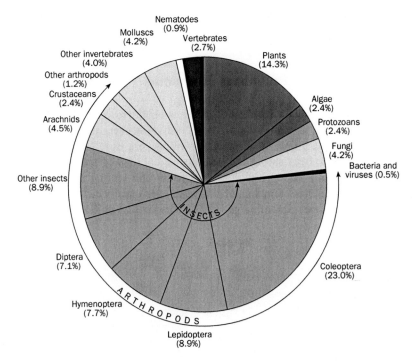

Figure 3.4 Proportions of different taxa from described species. The diagram includes taxa likely to contain 100,000+ species plus vertebrates

Source: Redrawn from Groombridge (1992).

Estimates of known species currently range between 1.4 million and 1.7 million species. The Global Biodiversity Review (1992) settles at a figure of 1.7 million described species, whilst UNEP's Global Biodiversity Assessment (1995) gives 1.75 million. But the species we know of are vastly outnumbered by those we do not. Those we do know are a biased sample, typically large, attractive taxa (mammals, butterflies), those that are easy to find, in thoroughly studied regions plus pests and parasites associated with any of the above (see Figure 3.4).

A major programme is now underway to describe all species on Earth and provide an encyclopaedia of life, in part relying on advances in information technology. In 2001 a summit of taxonomists agreed that a complete census was possible within twenty-five years (at an estimated $20 billion) and launched a three-part strategy. First the Catalogue of Life, to collate existing species data, building on existing 'Species 2000' and 'Integrated Taxonomic Information Systems' projects. Second, an initiative to describe all other species, spearheaded by the All Species Foundation in the United States. Finally an electronic Encyclopaedia of Life, the definitive guide to biodiversity from genome to ecosystem. The work has attracted sponsorship to the tune of millions of dollars and pulled together the Global Taxonomy Initiative (a product of the 1992 Rio Summit) and the Global Biodiversity Information Facility, an **OECD** (Organization for Economic Cooperation and Development) offshoot. Other scientists remain sceptical and many in the developing world are suspicious that the catalogue could be used to monopolise and exploit information, which they cannot effectively access. Nonetheless in 2025 we may be able to call up every life form on Earth, at the click of a button. In the mean time you can help. Biopat, a German company, is funding research into new species and for a donation you can have a creature named after you, the donation going to fund research in the field and lab. A Madagascan 'Greenish Boophis with red spots' tree frog could be yours for $5600, or a 'Creative Beatle' from Papua New Guinea for about $3000. Estimating this likely total is as important and as fraught as totalling up numbers of described species.

Estimating the actual numbers of species

If we cannot agree on the numbers of species we claim to know, the chances of estimating numbers of those we do not know seems rash. However, the actual total of species on Earth at the present time has become something of a Holy Grail in recent years. Many approaches have been used to estimate the actual total of species. Most involve using some data from a taxon, region or ecosystem and scaling up the patterns to global proportions. Each has strengths and weaknesses and usually different answers. The *Global Biodiversity* review (Groombridge 1992) settles for 12.5 million species (see Figure 3.5).

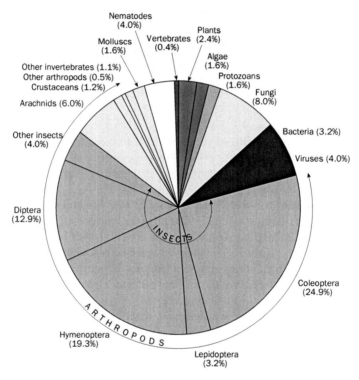

Figure 3.5 Proportions of different taxa from conservative estimates of actual total numbers of species. The diagram includes taxa likely to contain 100,000+ species plus vertebrates

Source: Redrawn from Groombridge (1992).

Time series

This technique uses the rates at which new species are being described and extrapolates the trends to look for an asymptote. Whilst we may be reaching an asymptote for a few taxa (e.g. birds and mammals) there have been recent bursts of finds for many invertebrate groups so that time series seem very unreliable (see Figure 3.6).

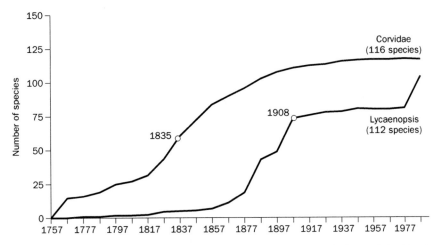

Figure 3.6 Numbers of described Corvidae (crows) and Lycaenopsis (part of the blue butterfly family). The dates highlighted indicate points at which half the species known in 1987 were recorded. Note the limited recent additions to Crow list but bursts of additions, some recent, to butterflies

Source: Redrawn from Groombridge (1992).

Expert opinion

Specialists with intimate knowledge of particular taxa can be asked to provide their best guess of likely totals and the results summed. Expert opinion has varied greatly over time, with estimates being revised ever upward. In the Seventeenth century John Ray thought there were between 10,000 and 20,000, while Hutchinson (1959) put the figure at 1 million.

Empirical relationships

There is a host of ecological patterns, often described in detail for individual communities, that can be scaled up from the original local data to global proportions (see Box 12). These include the ratios of species number to body size, species to area and species to population abundance, plus patterns derived from food webs and plant–herbivore associations (see Figure 3.7).

Taxon to taxon, region to region ratios

Such techniques rely on possession of good data for the numbers of taxa (both well-known and more obscure groups) in a particular region. The ratio of species of a well-known taxa (e.g. butterflies) to total taxa (e.g. all other insects) for the reliably surveyed region can then be used to estimate total numbers by counting the numbers of the same well-known taxa in comparatively poorly surveyed regions, assuming the same ratio of well-known taxon species to all other species holds true and multiplying up to get an estimate of the total. A classic example has been the use of butterfly to insect ratios from the United Kingdom. British insect fauna is unusually well recorded, so the data are reliable. There are 67 species of extant butterfly to about 22,000 insect species in total, a ratio of 1:328. Globally there are some 17,500 species of butterflies. Assuming the UK butterfly to total insect ratio holds globally, this gives an estimate for insect species numbers of 5.75 million.

Box 12

Erwin's famous 30 million species of arthropod

In 1982 entomologist Terry Erwin published a famed minimum estimate of arthropod species numbers, an astonishing 30 million. Erwin's estimate was based on a study of the insect fauna found on *Luehea seemannii*, a canopy tree, in Panama. He used insecticide fog, sprayed into the tree to knock down insects. He caught some 1200 species of beetle, of which 163 were specific to the *L. seemannii* tree. He then used a series of assumptions to reach a global estimate of tropical forest insect species richness. First, that there are 50,000 tropical forest tree species and that each would harbour 163 specialist beetle species, i.e. 8 million beetles. Second, that canopy beetles represented 40 per cent of canopy arthropods so that the total of canopy arthropods was 20 million. Finally, that there were twice as many canopy species as ground species, so add another 10 million to reach the final 30 million.

Erwin's assumptions are rightfully open to criticism but the very magnitude of the estimate was enough to prompt resurgent interest in the inventory of life on Earth. Erwin's approach has been revisited and refined with improved data for the assumptions, such as how many host-specific species of insect occur on each tree species and incorporating other plants such as the epiphytes that grow perched on trees. These results give estimates of between 2.4 million and 10.2 million arthropods, in line with estimates using other techniques such as body size or geographical patterns, which lie between 2.75 million and 10 million.

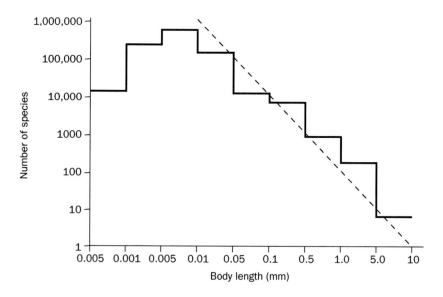

Figure 3.7 May's 'crude estimate' of the abundance of terrestrial animals of different body sizes. The dashed line is a similar model for the relationship based on the work of Hutchinson and MacArthur. Patterns such as these can be used as the basis for estimating total species diversity, extrapolating from what is known to include poorly described taxa

Source: Redrawn from May (1986) *Ecology*, 67.

Problems with estimation techniques

Scaling up

Many of the methods used rely on taking detailed data from a local, thoroughly documented site and scaling up to global proportions. Very many ecological patterns and processes are scale dependent. The patterns shown in local studies might not scale up globally in a reliable way. Patterns and processes could be diminished or exaggerated.

Comparing taxa

These gross estimates commonly lump taxa together. Yet this assumes that patterns and processes that work well for one taxa might not fit another. Data for ants might not be reliably extrapolated to estimate beetle numbers.

The fraction of undescribed species

Estimates rely on what we already know. Data used as the basis of a model may still be missing many undescribed species. It is a catch-22 situation: to estimate numbers of undescribed species we have to know about the undescribed numbers in the source data we use (see Box 13).

Species definitions

The criteria used to define a good species vary with taxa. The minute separations used to distinguish sub and sibling species amongst the birds are not applied to the desperately underworked micro-organisms. These problems afflict habitats differently. Sub and sibling species separations are poorly drawn in marine studies compared to land.

Box 13

Soil fauna: the other last biotic frontier

Deep-sea species diversity has been cited as the last frontier that is largely uncharted. Recent data of soil microarthropods from the soils of a typically harsh but accessible habitats sand dunes have suggested soil fauna deserve an equal billing. Samples yielded densities of up to 1.4 million individuals per square metre, compared to previous studies of soil fauna from as low as 3000 per square metre in Indonesian rainforest, through 110,000 in Congo rainforest litter to 386,000 per square metre in Alpine grassland. The numbers of species were small, only 31, but mostly undescribed. More important was the small size of many of the new species, less than 0.2 mm long. Many classic estimates of total species numbers rely on extrapolating known patterns of body size relative to abundance from local studies up to a global scale. The abundance of small-bodied species from this one study of dunes would be enough to increase Robert May's classic 1988 estimate of 10 million terrestrial animal species, based on body sizes down to 0.2 mm, up to 20 million.

The workforce

The size, expertise and where in the world the workforce is based all affect estimates.

The major gaps

The problems of bias amongst described taxa, estimation techniques and expertise have left major gaps in our knowledge of biodiversity, including particular taxa, habitats and conceptual topics, super-rich groups and reference sites.

Ecosystems

Marine systems

Although marine systems cover most of the planet, our knowledge is very patchy. Whilst the biodiversity associated with coral reefs has been compared to that of tropical rainforests, other habitats such as the surface waters of open oceans resemble deserts. Sampling of deep sea sediments coupled with greater interest in micro-organisms has provided evidence of very high diversity. The diversity of deep sea life may be a result of the great extent of this habitat, a macro-ecological phenomenon if local diversity is sustained by the extended regional species pool. The open ocean can spring surprises. In 1988 a new genus of bacterium, *Prochlorococcus*, was identified; it may be the most abundant cellular life on Earth.

Tropical rainforests

Although rainforests appear to support the greatest biodiversity on Earth, research effort is still overshadowed by the sheer size of the task. Particular habitats within the forest have yielded tantalisingly diverse communities, but whether such results can be extrapolated reliably remains unknown. For example, one explanation for the variety of herbivorous tropical forest insects is that each is very specialised, the forest environment in effect parcelled up into very small niches, each species perhaps feeding on only one host plant. Recent results suggest that these insects may be less specialised, with perhaps only 2–5 beetles unique to each tree species, rather than the 160+ used to estimate Earth's arthropod richness at 30 million (see Box 12).

Intraterrestrials

Up until the 1980s it was believed that life was restricted to the upper metre or two of soil or seabed. Revelatory discoveries have now detected life in rocks as deep as 3.2 km down in South African gold mines, with some microbiologists suggesting we live on top of a 6 km deep, hot biosphere. These crust-dwelling microbes have been named the intraterrestrials. Some appear to rely on energy sources derived from the rock itself, for example water–basalt reactions that release hydrogen, although other communities may be relying on percolation of dissolved organics. DNA sequences from marine and terrestrial samples suggest many are related to the microbes of deep sea hydrothermal vents. The rocks may also be old. Intraterrestrials have been detected in 14 million year old marine sediments. Intraterrestrial life appears to go very slowly, effectively dormant for long periods and is lived at low densities. These extraordinary communities remain elusive.

Taxa

Parasites

Given that most organisms seem to suffer parasitism, it is tempting to imagine parasite diversity as vast as the range of potential hosts. This is especially so in the case of those insects that lay their eggs in or on living hosts (typically other arthropods), the host being killed only once the parasite matures and hatches. Such insects are specifically termed parasitoids; this distinction from other parasites reflects the almost invariably fatal impact on the host, unlike other parasites. Given that insects make up the overwhelming majority of species, if every species had its own parasitoid (and parasitoids are in their turn attacked by their own super-parasitoids) this automatically increases total species massively. However, evidence of parasite and parasitoid biodiversity is sketchy.

Micro-organisms and fungi

Diversities of microbial eukaryotes, Archea and Bacteria, Fungi and viruses are poorly known, despite the significant, perhaps dominant, roles they play in the Earth's geochemical cycles. Terrestrial subsurface microbial biomass may rival that of all land and marine plants whilst oceanic viruses are the most abundant of all life forms. Difficulties of sampling, identification and defining exactly what is a species plague the work with these groups. Free-living microbial eukaryotes (often more familiar as 'Protozoa') may be relatively species poor, with estimates of between 10,000 and 20,000 species. Most species have global distributions and an abundant seed bank of cysts or spores allows species to reappear when environmental conditions suit. They show no geographically driven diversity, no known endemics. The diversity of prokaryotes, Archea and Bacteria is difficult to estimate because clonal reproduction and the frequency of gene transfer undermine the use of definitions of species familiar to animals and plants: 4500 have been described but analysis of prokaryote genetic diversity in soil and water samples suggests this is wholly unrealistic. DNA-based techniques define species as genomes with 70 per cent or more homology of DNA. Using this criterion, soils and water samples of between 30 and 100 cubic centimetres contain between 3500 and 11,400 genomic species. There could be millions of prokaryotic species. The visible wildlife we hold dear, the giant pandas and blue whales, are no more than surface decoration.

Nematodes

Nematodes are a phylum of worms, capitalising on a very uniform body plan but extremely abundant in very many habitats. Problems from lack of expertise and identification have hindered investigation of this phylum but recent evidence hints at high species diversity. About 15,000 species are known but some nematode experts have suggested that there may be millions of species, basing this claim on the abundance of these worms. As the apocryphal story puts it, they are numerous and found throughout all habitats and organisms so that if all other organic life was to be stripped away a ghostly outline would remain sculpted in nematodes.

Mites

These are members of the Phylum Arthropoda. Like the nematodes, problems of lack of expertise and identification have held back work but there are suggestions of high species diversity.

Insects

Despite the considerable work devoted to the class Insecta there remain very poorly known groups and habitats. Many of the estimates of total global species richness are based on insect or arthropod richness (see Box 12).

Conceptual gaps

Super diverse groups

Some taxa have been described as super (or hyper, or ultra) diverse (or speciose), reflecting an inordinate variety of species. This accolade suggests that the patterns and underlying processes may be unusually magnified or even different to the typical determinants of diversity. Understanding such differences is not only important but also difficult and may require separate models to estimate total numbers.

Special species and areas

All diversity is not equal. An individual species, perhaps just one comprising an unusual lineage, might be rated as special, more worthy of protection than another species with many related species in a diverse genus. How to measure the phylogenetic isolation and uniqueness of a species is a new field. Some taxa may be very useful indicators of the overall diversity of a site, their diversity accurately reflecting the variety of other species which cannot be measured for want of time and techniques. Similarly focal taxa are groups combining concepts of the special and representative and could be useful signals of important habitats. Essential areas and reference sites are also concepts only recently developed. The idea of **centres of diversity** has been developed, reflecting the evolutionary heartland (or perhaps last refuge) for a taxon. Linked to this, **complementarity** has been developed as one criterion to assess regional diversity, especially for plants. All the species of a taxon make up the total complement. The single most important site (or country) for this taxon is that with the highest proportion of this complement; the site with the highest proportion of the remaining species becomes the next most valuable. Conversely centres of diversity for birds have concentrated on aggregations of endemic species. Regions of high diversity or endemicity for one taxon are not necessarily so for another. All these ideas, their usefulness and identification are comparatively new, debatable and poorly researched.

Global biodiversity patterns

Taxonomic biodiversity: global patterns of taxonomic diversity

Global patterns of species diversity are well known for only a few groups. There are marked patterns, in particular concentrations of species within biodiversity hot spots, biogeographical variations between continents and trends such as the increase of species in most taxa towards the tropics. Birds and plants have been studied in detail, reflecting a bias in expertise and available data, but already producing useful insights for conservation planning.

Bird species number nearly 9700, the most speciose of terrestrial vertebrates compared to over 4000 amphibians, 6550 reptiles and 4327 mammals. Birds are one of the most thoroughly recorded taxa: their distribution is unevenly spread across the main biogeographic realms, with 3083 species in South America, 2280 in Asia, 1900 Africa, 950 in the Palaearctic, 900 in Australiasia and 800 in the Nearctic. Within continents the uneven pattern continues. Table 3.5 outlines species, family and endemic diversity from the top five South American countries plus Galapagos Islands (see Figure 3.8).

Birds have been used as a focus to identify priority areas for conservation, primarily using endemics. Birds are potential good indicators of biodiversity. Birds are found throughout the world and have diversified in all terrestrial regions and habitats. Their

Table 3.5 Bird taxonomic diversity of top five South American countries plus Galapagos Islands. Note Ecuador's high diversity despite small size and Galapagos' proportion of endemic species

Country	Area, km²	Species total	Endemic species	Family total	Endemic families	Vascular plants, species numbers
Colombia	1,138,914	1,700+	67	79	20	51,220
Peru	1,285,216	1,538	112	86	23	17,144
Brazil	8,511,965	1,492	185	84	23	56,215
Ecuador	283,561	1,388	37	82	21	19,362
Venezuela	1,098,581	1,340	40	79	20	21,073
Galapagos Islands	7,845	136	25	38	0	

Sources: Groombridge 1992; Wheatley 1994; Heywood 1995; Groombridge and Jenkins 2002

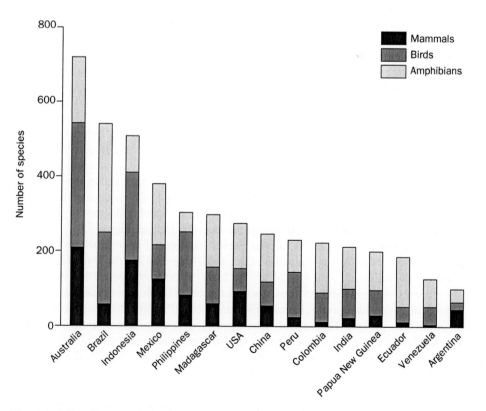

Figure 3.8 Top fifteen countries for endemic mammals, birds and amphibians

Source: Data from Groombridge (1992).

Plate 13 Mount Kupe, part of the Cameroon Mountain Endemic Bird Area.

taxonomy and species distributions are well known; there is also evidence for correlation between their diversity and that of other terrestrial vertebrates and vascular plants. The International Council for Bird Preservation has run an extensive global programme to identify **Endemic Bird Areas (EBAs)**. The survey relies on endemics, or more exactly restricted range species breeding over a total range of no more than 50,000 sq. km (see Plate 13). Any areas with two or more restricted range species are cited as EBAs; 2609 birds fitted the range criterion, plus 59 known to have gone extinct since 1800. The analyses suggested 221 EBAs, 70 per cent in the tropics, embracing 2480 species, 77 per cent of known threatened bird species. EBAs are ranked into three categories: Priority I is most important/threatened while Priorities II and III are less important/threatened. These categories include estimation of biological importance (restricted range species relative to area, taxonomic uniqueness and known significance to other plant and animal taxa) and degree of threat (threat to birds, proportion of area within IUCN recognised **protected areas**). Table 3.6 gives numbers of EBAs, bird species and numbers in priority categories for the five countries containing the largest number of EBAs (see Figure 3.9).

All five countries of Table 3.6 harbour tropical rainforest. For comparison Table 3.7 provides details of EBAs for Zimbabwe, Cameroon, Mauritius and the United Kingdom.

Bird species diversity patterns can now be analysed using extensive DNA databases. The evidence suggests that many endemicity hot spots in South America and Africa may be a mix of ancient lineages that have survived and recent (Pliocene/Pleistocene) radiations of other bird groups. The habitat heterogeneity and disturbance in the hot spots simultaneously allow some relict species to cling on in some patches and also act as a spur to radiations in adjacent habitats. These species factories are often found in the zones between major biomes (see Figure 3.10).

Plant species diversity is unevenly divided between four main Phyla: the Bryophytes (mosses and liverworts) 16,000 species; Pteridophytes (ferns and allies) 10,000–12000 species; Coniferophytes 700 (conifers, often called by the slightly older name Gymnosperms); Anthophytes, 250000, perhaps up to 750,000 (flowering plants, often

Table 3.6 Five countries with largest numbers of Endemic Bird Areas

Country	EBAs (inclusive of those shared with other countries)	Restricted range species confined to countries' EBAs	Total restricted range species, confined plus shared	Priority I EBAs	Priority II EBAs	Priority III EBAs
Indonesia	24	339	411	16	7	1
Brazil	19	122	201	8	8	3
Peru	18	106	216	4	7	7
Colombia	14	61	189	4	9	1
Papua New Guinea	12	82	172	3	63	3

Source: Bibby 1992

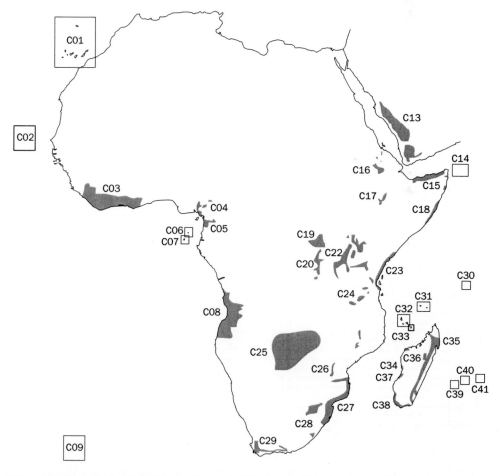

Figure 3.9 African Endemic Bird Areas. The prefix 'C' refers to EBAs from Africa, Europe and the Middle East under the International Council for Bird Preservation EBA scheme. Mount Kupe (Chapter 4) is part of the Cameroon Mountain EBA, C04, Mauritius, once home to the Dodo, C40 and Zimbabwe harbours C26, the east Zimbabwean Mountain EBA

Source: Redrawn from Bibby (1992).

Table 3.7 Endemic Bird Areas of Zimbabwe, Cameroon, Mauritius and UK

Country and name of EBA	Area of EBA, km²	Restricted range species confined to EBA (+ shared with other EBA)	Per cent of EBA area protected	Priority category
Zimbabwe East Zimbabwe Mountains	4,900	2 (+ 2)	5–10	III
Cameroon Cameroon Mountains	7,300	26 (+ 2)	5–10	I
Cameroon Cameroon/ Gabon Lowlands	40,000	5 (+ 1)	20–30	III
Mauritius Whole island system	1,900	6 (+ 4)	0–5	I
UK No EBAs	—	1	—	—

Source: Bibby 1992

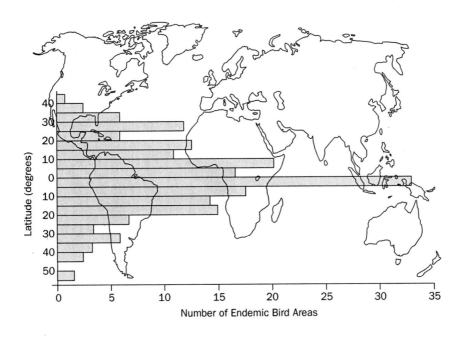

Figure 3.10 Latitudinal gradient of numbers of Endemic Bird Areas

Source: Redrawn from Bibby (1992).

called Angiosperms). Pteridophytes, Coniferophytes and Anthophytes are often separated out from Bryophytes as the vascular plants, defined by possession of effective water transporting tissues. Diversity within the three vascular plant groups is biased towards some families. Anthophyte diversity is dominated by a few families, e.g. Orchidaceae (orchids)with between 25,000 and 35,000 species, Compositae (thistles, dandelions) 20,000 and legumes (peas and beans) 14,200. The Coniferophytes contain 500 conifers, 100 cycads then some numerically tiny but remarkable species, notably the fossil relict Gingko tree (one species) and the slow-growing, two-leafed *Welwitschia* of southern African deserts. The Pteridophytes consist of 9000 fern species and again, small numbers of allied groups including horsetails and Lycopods, plants that were once the dominant vegetation of Palaeozoic forests growing as trees. Plants are unevenly distributed between the continents: Latin America 85,000 species, Africa 40,000–45,000, Asia 50,000, Australia 15,000, North America 17,000 and Europe 12,500. Eighteen plant diversity hot spots support 50,000 species, 20 per cent of the world's vascular flora, but between them cover only 5 per cent of the land's area.

Species richness within the sometimes unnatural confines of political boundaries reflects size, topographic heterogeneity and complexity plus climate. Detailed studies of tree and vine inventories from South American tropical forests have compared species richness to climatic factors such as total rainfall and seasonality, plus soil nutrients. Diversity showed clear increases with total rainfall and length of rainy season. Although climate alone was a very effective predictor of species richness, soil factors could be important, especially increased species richness with increased nitrogen availability at higher altitudes. Such results can be useful to focus attention on known high total rainfall and long rainy season sites as of particular importance for conservation, even if good species inventories are not yet available. The greatest species richness is in the tropical forests but many oceanic islands have very high endemicity. Tiny St Helena in the Atlantic, Napoleon's final exile, has seventy-four endemics out of eighty-nine species. Table 3.8 gives known vascular plant numbers and endemicity for the top five species-rich countries and other countries for comparison (see Figure 3.11).

Revelations of plant species richness continue at a smaller scale. A study of a one hectare plot in the Ecuador *terre firme* tropical forest in the late 1980s found 473 species (with stem diameter at standard breast height of greater than 5 cm). Given the estimated 3000 tree species in Ecuador, this represented 16 per cent of the country's tree flora but 49 per

Table 3.8 Flowering plant diversity, selected countries

Country	Flowering plant species	Conifers and allies	Ferns and allies	Number of endemics (% of species)	
Brazil	55,000	—	—	—	
Colombia	35,000	—	—	1,500	(4.3)
China	30,000	200	2,000	18,000	(56)
Mexico	20,000–30,000	71	1,000	3,624	(13.9)
USSR (as was)	22,000	74	207	—	
Madagascar	8,000–10,000	5	500	Up to 8,500	(68)
UK	1,550	3	70	16	(1)
Zimbabwe	4,200	6	234	95	(2.1)

Source: Groombridge 1992; Heywood 1995

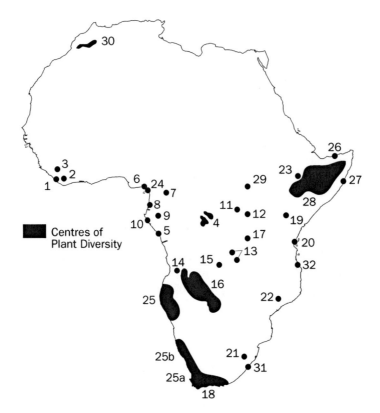

Figure 3.11 African Centres of Plant Diversity identified by IUCN. Sites are selected due to particular richness of flora which, if protected, would safeguard the majority of the earth's wild plants

Source: Redrawn from Groombridge (1992).

cent of species were represented by only one individual. Similar inventory work from Colombian forest, working in even smaller plots of 0.1 hectare area but including trees and herbs, found up to 313 species per plot.

Plant species diversity has also been the target of the IUCN Plant Conservation Programme to identify **Centres of Plant Diversity** (CPDs). The rationale for CPDs is that their identification and subsequent protection will conserve the majority of wild plants; 241 CPDs have been identified so far. The main criteria are species richness and numbers of endemics. The IUCN programme also takes into account value of the gene pool to humans, diversity of habitat types, sites containing significant proportion of species adapted to specific conditions found therein and imminent large-scale threat.

Ecological biodiversity

Describing the extent of global ecosystems

Ecological biodiversity shows global patterns akin to taxonomic trends. Global classifications depict polar to equatorial biomes, largely created and defined by geography and climate. Some habitats are restricted by latitude (e.g. tropical rainforest, coral reefs) but others occur across the planet, though their precise regional form and complement of species varies.

Global biogeography has long recognised differences between the major continents and their allied peripheries (see Figure 3.12). The six main divisions are Nearctic (North America), Neotropical (Central and South America), Palaearctic (Europe, Russia, central and eastern Asia), African (Africa including Madagascar), Oriental (India, Indo-China and South East Asian islands down to Borneo) and Australasia (Australia, New Zealand, New Guinea).

Increasingly detailed global classifications of biomes have been devised, increasingly using remote sensing from satellites. These attempt to classify the main terrestrial ecosystem types produce schemes that broadly reflect global climate systems and sometimes other natural or human factors, e.g. soils or landuse. Udvardy's (1975) combination of the use

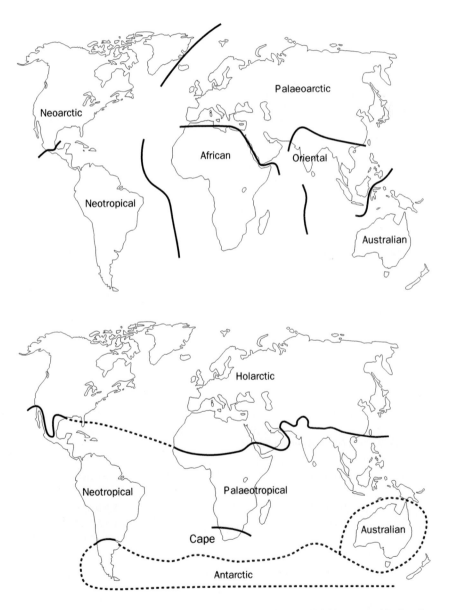

Figure 3.12 The classical animal (above) and plant (below) global biogeographical realms

Source: Redrawn from Heywood (1995).

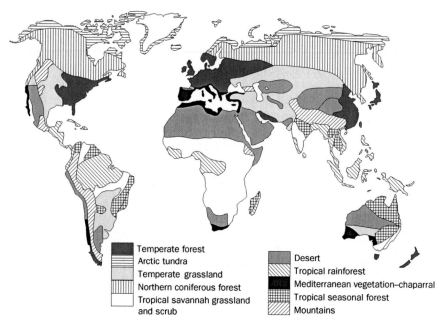

Figure 3.13 Gross global vegetation classification

Source: Redrawn from Heywood (1995).

Table 3.9 Example vegetation types and extent from Udvardy's system

Realm	Biome and province	Area, km²
Afrotropical	**Tropical Humid Forest**	
	Congo rainforest	2,195,019
	Guinean rainforest	709,112
	Malagasy rainforest	147,862
	Evergreen Sclerophyllous vegetation	
	Cape Sclerophyll	99,663
Neotropical	**Tropical Humid Forest**	
	Amazonian	2,864,623
	Campechean	279,695
	Colombian coastal	273,266
	Guyanan	1,090,396
	Mudieran	1,988,840
	Panamanian	128,872
	Atlantic forest	223,944
	Evergreen Sclerophyllous vegetation	
	Chilean sclerophyll	47,988

Source: Hannah *et al.* 1995

of dominant vegetation type and biogeographic is still widely used and has been used as a base against which to measure human impacts on biomes at a global scale (see Figure 3.13).

Udvardy's (1975) global classification is still widely used, combining biogeographical and biome-based categories. Udvardy based his system on eight major biogeographic realms: Afrotropical, Antarctic, Australian, Indo-Malayan, Nearctic, Neotropical, Oceanian and Palaearctic. Major vegetation types, biomes, were then assessed within each. The biomes were identified as provinces reflecting regional variations in form. Table 3.9 lists Afrotropical and Neotropical examples.

The distribution of wetlands varies globally, regionally and nationally. The extent of global wetlands has been estimated from compilations of maps and remotely sensed images from satellites. Estimates of the global area of wetlands vary with definitions and inclusion of coastal wetlands but range between 5.3 million and 8.6 million sq. km, compared to estimates for extant tropical rainforest of 9.4 million to 12 million sq. km and for grasslands of 24 million to 35 million sq. km. The total extent of wetlands varies with latitude and the different types of wetland are unevenly distributed with vast tracts of bogs dominant across northern temperate and subarctic continents versus the swamps and floodplains of the tropics

The distribution varies with scale. Table 3.10 compares wetland distributions globally and in Europe. Differences would affect assessment of conservation priorities (see Figure 3.14).

Table 3.10 Extent of freshwater wetlands, globally and in Europe versus Africa

	Bogs	Fens	Swamps	Marsh	Floodplain	Lakes	Mangrove	Anthropogenic
Global	1,867	1,483	1,130	274	823	114	27	1,300 (rice paddies)
Europe	54	93	1	4	1	1	0	Minor
Africa	0–38		85	57	174	39	6	46

Source: Groombridge 1992

The importance of global ecosystem: functional biodiversity

Central to the concept of biodiversity is the sense that ecosystems are important for what they do. Ecosystems carry out functions, thereby providing services. Ecosystems are responsible for fluxes of energy and materials. The importance of ecosystems as providers of services adds to our awareness of the dangers from degradation and loss and of the value of biodiversity. Functional biodiversity meshes the concept of the ecosystem, dominated by ideas of flow and flux of resources with the community, the numbers of species. Ecosystems are a synergy of genetic, population, community, ecosystems and landscape biodiversity. Ecosystem function is the sum total of their activity, apparent even when the precise importance of individual species as keystones, driving the processes remain unclear. The array of services include vital global life support through atmospheric quality, climatic control, protection of coasts and nutrient cycling. The importance of functional biodiversity is best revealed by an example of just one ecosystem.

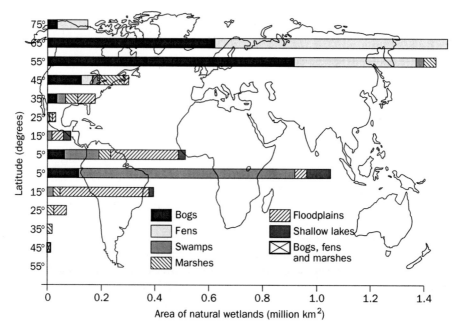

Figure 3.14 Latitudinal extent of different wetland types

Source: Redrawn from Aselmann and Crutzen (1989).

The importance of wetlands

Wetlands throughout the world support their own unique wildlife. By its very existence this biodiversity has an intrinsic value. Tangible, highly valuable benefits also derive from wetland ecosystems. These can be broadly divided into **production** (in ecological terms the living components, in economic terms the stock), the **ecosystem functions** (in economics, the services) and the **intrinsic cultural and biological diversity** in its own right (attributes to economists). The astonishing variety of benefits, local to global, provided by wetlands is outlined next.

Production

Wetlands are the most productive natural ecosystems on Earth. The following examples are all sustainable uses of wetlands, although in some cases more intensive versions of the same exploitation also exist, destroying the original habitat (see Plate 14).

- *Consumptive:* wetlands yield fish, shellfish and prawns. Plant products include wild fruit and vegetables, gardened crops and small-scale arable. In addition plants provide grazing plus harvested fodder, building poles, thatch, craft materials, latex, resin, tanning products, beer and medicines. Note that these products are not the preserve of subsistence economies. High vale crops such as thatch, rushes, sedges and wild cranberries are harvested in developed countries. Animals are hunted for fur and skins, as well as sport. Energy is provided by fuelwood and peat.
- *Non-consumptive:* tourism and recreation rely on the wetland landscape and wildlife without being direct consumers (see Plate 15).

Plate 14 The value of wetlands: direct use production. Saw sedge (*Cladium mariscus*) harvested at Wicken Fen, Cambridgeshire. The sedge is used to protect peaks and angles of thatched roofs and is a high value crop.

Plate 15 The value of wetlands: ecosystem services. The Insh marshes in Scotland, a Royal Society for the Protection of Birds reserve, act as a buffer, holding flood waters that spill out over the marshes from the River Tay, helping to protect land downstream from flooding.

Services

- *Storage:* wetlands act as major water stores, allowing ground water to recharge as water slowly seeps into deep aquifers and also to discharge, as ground water tops up surface aquatic habitats such as rivers and lakes. Note that wetlands can act as both recharge and discharge sources, perhaps switching roles with the seasons. Even small wetlands can act as insurance supplies during drought. Trapped sediment can build up land. Peat bogs are major sinks for carbon dioxide because peat is organic (primarily plant) material that cannot completely decay so that the carbon is not recycled.
- *Buffering:* this is the ability to slow, compensate and ameliorate against change. Wetlands buffer many potentially destructive environmental processes, reducing both the size and rate of change. Coastal wetlands provide shore defences, dissipating wave and storm energy. Wetlands act as flood overspill areas holding water that may do damage elsewhere, slowing the rise of floodwaters and desynchronising floodwater peaks from different rivers which might otherwise combine. Climate is also buffered so that in hot weather, wetlands act to cool the local climate, and in cold weather warm up their environs. This benefit has been used in fruit farms in the United States to guard against sudden cold snaps.
- *Cleaning:* wetlands are effective filters, particularly between rivers and their surrounding catchments. Materials are trapped and held, the wetlands acting as sinks, and rates of flow are slowed allowing extra time for natural processes to neutralise potentially harmful inputs. Wetlands are known to trap and clean sediment run-off, organic detritus, sewage, excess nutrients (especially phosphorous and nitrogen compounds), acidic inputs, metals, pesticides and pathogens. These effluents can be from a readily identifiable point source such as a mine-working or sewage treatment plant. In addition wetlands are particularly useful to control diffuse, non-point source inputs that plague so many catchments, e.g. nutrient run-off from farmland, which cannot easily be collected and cleaned by technical intervention.
- *Pathways:* the waterways ramifying throughout many wetlands provide pathways for natural fluxes such as nutrient recycling and alluvial deposition, increasing fertility. Open water also provides literal routes for movement by animals and humans.

Attributes

Cultures utilising wetlands have developed customs, architecture and landscapes as a result of this interdependence. In some cases the entire society is intimately tied to the wetland system, creating unique cultural systems, e.g. the Marsh Arabs.

So often dismissed as miasmal wastes in urgent need of draining, wetlands are actually a vital asset (see Box 14).

Islands and hot spots: the Earth's extra-special places

Biodiversity is not evenly distributed. Within the general patterns of taxonomic and ecosystem distribution, there are striking concentrations of species, associated with certain ecosystems. Identification of these special places is important for practical conservation and to understand the underlying causes that drive diversification. These foci can be special by virtue of several, sometimes contrasting, factors:

- very high total species numbers
- endemicity, whether of common or unusual lineages
- unusual combinations, characteristics of communities
- super speciose taxa.

Box 14

The role of wetlands: global and national ecosystem services

Global: methane production

Methane (CH_4) is a greenhouse gas that is also important in the formation of ozone. Concern at possible global warming has prompted detailed analysis of global biogeochemical cycles, including methane production which arises from natural sources such as digestive processes of ruminant animals such as cows and decay of organic matter, plus human sources such as use of fossil fuels and burning. Atmospheric methane is increasing by about 1 per cent per year, i.e. 40–50 Teragrams (1 Teragram = 10^9 kg) which, allowing for the balance between production, recycling and use, requires about 400–600 Teragrams to be pumped into the atmosphere annually. Estimates of methane production from wetlands are 40–160 Teragrams per year from natural wetlands plus 60–140 Teragrams from rice paddies. The most important regions are northern latitude fens 50–70°N, subtropical paddies 0–20°N and southern hemisphere tropical swamps 0–10°S. Although methane emissions from wetlands are highly variable depending on seasonal temperature and waterlogging, and estimates of global methane fluxes are still tentative, the suggested mean methane emissions from wetlands may represent up to half the global annual production of methane and are therefore very important for planetary health.

National: dambo horticulture in Zimbabwe

Small, seasonally inundated valley wetlands, sometimes herbaceous, sometimes wooded, are found in the headwaters of drainage basins throughout southern Africa. Zimbabwe, although primarily an arid country, is particularly rich in these ecosystems, known by the local name of dambos. Zimbabwean dambos have been used since at least the Iron Age as part of a shifting system of cultivation and grazing. Dambos are fertile and remain moist even in years with poor rainfall so are an especially valuable safety net in an uncertain world. Dambo wetlands are primarily used as small (0.1–2 hectare) market gardens. The fertility and water supply permit diverse crops, from staples such as maize and rice to water-greedy pumpkins and fruit, which can be cropped all year. Limited grazing and beekeeping are possible. Dambos also act as water sources for drinking, rivers, irrigation and livestock during drought and as water stores, mopping up excess during rare floods. Dambos are a special asset to the rural poor as the horticultural garden produce is a valuable source of income. The nature of this local, small-scale horticulture is a valuable economic opportunity for rural women. During recent drought years 80 per cent of households with access to market gardens on dambo wetlands remained self-sufficient in food (see Figure 3.15).

These jewels in biodiversity's crown fall into three contrasting types. First **continental hot spots** sites of very high diversity, often with unusual endemics too, sometimes called mega, hyper and super diversity centres. Second, **large islands** (sometimes called **continental islands**) harbouring diverse, distinctive faunas which include relict fauna long extinct on the main continents. Finally, there are small **oceanic islands**, often low in total

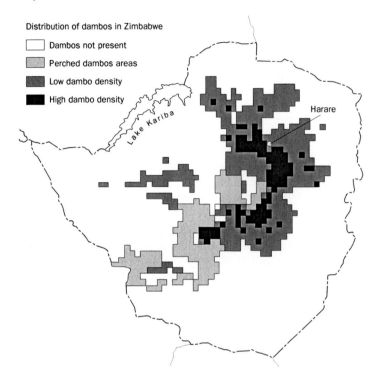

Figure 3.15 Extent of dambos in Zimbabwe

species numbers (though a few individual taxa can be unusually speciose) but with high proportions of endemics, unusual combinations of species, and peculiar evolutionary lineages. Merely listing high diversity centres by country is possible but not very informative since many, especially the continental centres, do not occupy the whole of a country and some straddle borders.

Continental hot spots

Biodiversity hot spots defined by endemism and the uneven distribution of species have been cited for birds and plants in earlier sections. A few countries, primarily tropical, have been described as megadiversity countries, unusually rich in all forms of biodiversity, although data for such categorisation rely on higher vertebrates, plants and a few insect groups. The megadiversity countries are Mexico, Colombia, Ecuador, Peru, Brazil, Democratic Republic of the Congo (formerly Zaire), Madagascar, China, India, Malaysia, Indonesia and Australia. However, political boundaries are an inappropriate guide and generalisations do not take into account the uniqueness and biodisparity found in other countries. Eighteen global hot spots identified by endemicity of many taxa have been identified. Table 3.11 summarises data for four.

Globally significant continental hotspots are often regions of habitat heterogeneity caused by habitat change. The resultant patchwork allows older taxa to survive and radiation of new species (see Figures 3.16 and 3.17).

Islands

Island diversity centres are very different from the continental hot spots, often poor in species but sheltering unusual relicts, strange combinations and extraordinary evolutionary

Table 3.11 Example of continental biodiversity hotspots

	Higher plant species	Endemic mammal species	Endemic reptile species	Endemic amphibian species	Endemic swallowtail butterfly species
Cape region, South Africa	6,000	15	43	23	0
Colombian Choco	2,500	8	137	111	0
Western Ghats, India	1,600	7	91	84	5
South West Australia	2,830	10	25	22	0

Source: Bibby 1992

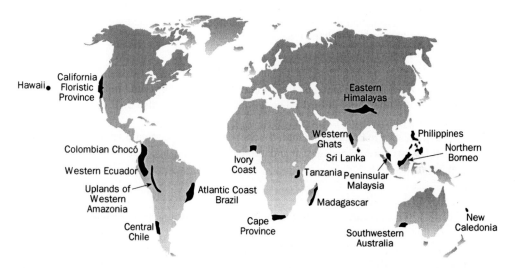

Figure 3.16 Global biodiversity hotspots, defined by endemicity and also threat from human pressures, identified by Wilson 1992

Source: Redrawn from Wilson (1992).

radiations. Islands also have a historic role in our awareness and understanding of biodiversity, be it the evolutionary inspiration of the Galapagos, or that icon of extinction, the dodo, from Mauritius.

Islands have distinctive ecological communities. Historical contingency (what was living there before islands split from mainlands, when the separation occurred) and the vagaries of colonisation create unusual combinations of wildlife. Relict faunas, long extinct on main continents, survive and in some cases flourish. In situ speciation creates spectacular radiations, such as the very speciose fruit flies of Hawaii. Animals take up unexpected roles; on Hawaii one moth has a caterpillar that does not chew leaves but is a deadly ambush predator, snatching other insects. A few taxa have become particularly characteristic of islands, e.g. pigeons, rails and tortoises, all capable of erratic but long-distance dispersal, yet not so mobile that once established, new arrivals dilute the evolutionary aftermath.

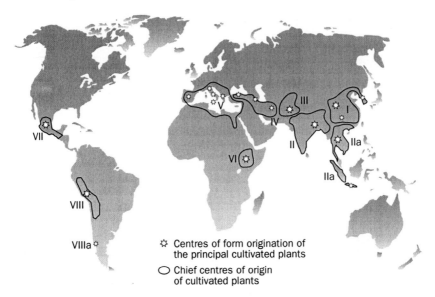

Figure 3.17 Centres of origin of cultivated plants, based on the work of Vavilov. Stars indicate the centres of origination for the form of crop plant used, the outlined surrounding areas the broad region from within which the plant was domesticated

Source: Redrawn from Heywood (1995).

The same traits develop amongst these taxa: gigantism in birds and reptiles, dwarfism in mammals, flightlessness in birds. The wildlife of islands is also distinguished by being extinction prone.

There are two main types of high diversity islands: continental islands such as New Zealand, Australia and Madagascar and tropical oceanic islands such as Hawaii, Galapagos and Mauritius.

Continental islands
Table 3.12 gives comparative biodiversity data for Australia, New Zealand, Madagascar and Kenya as a contrasting continental country.

Oceanic islands
Oceanic islands supporting special diversity are typically very isolated, volcanic (the altitude creating diverse habitats) and tropical. Low-lying tropical atolls lack habitat diversity whilst temperate islands are often too exposed to extremes of weather (see Table 3.13).

Island wildlife has proven particularly susceptible to extinction, due to several factors:

- *Evolutionary innocence*: island species have proven very vulnerable to introduced competitors, predators, parasites and diseases. Many islands lack taxa such as large mammalian predators and ants, the very groups that are dominant ecological keystone taxa on mainlands. Relict fauna and particular traits such as flightlessness further increase the threat posed by alien arrivals.
- *Small populations:* many island species occur as small, isolated populations specialised to live in a narrow habitat range and with no pool from which recolonisation can boost numbers. Genetic diversity may be enfeebled by inbreeding and bottlenecks due to population size. In addition co-evolution of island taxa one with another may result in extinctions cascading through the island once one or two species are lost and ecological links are broken.

Table 3.12 Diversity, threat and uniqueness of three continental islands, with Kenya as a comparison. Note that counts of extinct species exclude those lost from very small islands around the mainlands

	Australia		Madagascar		New Zealand		Kenya	
	Total	Endemic	Total	Endemic	Total	Endemic	Total	Endemic
Flowers	}		8,000–10,000		2,160		6,000	
Gymnosperms	15,000	80%	5	68%	22	82%	6	4%
Ferns			500		189		600	
Mammals	282	210	105	67	—	3	309	10
Birds	571	351	250	97	285	74	1,067	7
Reptiles	700	616	252	231	40	40	187	15
Amphibians	180	169	144	142	3	3	88	10
Fish	113	110	40	38	30	27		
IUCN Plant Centres of Diversity	8		1		3		1	
Endemic Bird Centres	7		5		2		2	
Threatened species								
Plants	2,024		194		232		144	
Mammals	38		50		1		17	
Birds	39		28		26		18	
Reptiles	9		10		1		2	
Amphibians	3		0		3		0	
Fish	16		0		2		0	
Animals known to have gone extinct	18		3		14		0	
Extinct endemic plants	173		0		5		0	
Unusual wildlife	Marsupial and reptile diversity		Lemurs; extinct flightless Elephant Bird		Flightless birds (Kiwi, extinct Moa)			

Source: Groombridge 1992

- *Lack of disturbance:* physical disturbance to island habitats can be increased once humans arrive, for example direct action such as fire to clear land or indirectly through impacts of introduced animals such as pigs grubbing up land.
- *Human exploitation:* direct human harvesting has wiped out species such as the dodo and moas and endangered others, e.g. giant tortoises. Habitat loss as humans clear land is just as great a threat, especially given the small habitat areas and species' ranges of many island taxa.

Islands remain under threat, even the most famous. The Galapagos Islands have become a focus of concern, in part due to increased tourism but also because of protests from local people angry at restrictions from conservation laws. Protests have culminated in intentional

Table 3.13 Diversity, threat and uniqueness of three oceanic island systems. Note that the data exclude introduced species

	Hawaiian Islands (Pacific)	Galapagos Islands (Pacific)	Mauritius (Indian Ocean)
Vascular plants	900 of which 850 are endemic	540 of which 170 are endemic	878 of which 329 are endemic
Threatened native plants	343	82	269
Introduced plant species	4,000	195	—
IUCN Plant centres of Diversity	1, whole archipelago	1, whole archipelago	1, whole archipelago
Endemic Bird Centres	1, whole archipelago	1, whole archipelago	1, whole archipelago
Animals known to have gone extinct	86	5	41
Extinct endemic plants	108	2	24
Species of land snails	c.1,000, all endemic	90 of which 66 are endemic	109 of which 77 are endemic
Extinct snail species	29	1	25
Unusual wildlife	High endemism, super-species rich taxa, e.g. fruit flies	High endemism, unusual species, e.g. marine iguana	High endemism; once home to the dodo

Sources: Stone and Stone 1989; Groombridge 1992

habitat destruction and threats to conservationists. The focus on hot spots has also received more theoretical criticism, suggesting that our fascination with these small areas risks not doing much to prevent the damage to the rest of the world, the cold spots, which provide the majority of the ecosystem functions upon which we rely.

More things in Heaven and Earth . . .

Fragments of ancient continents, far-flung islands and impenetrable jungle – our fascination with biodiversity is in part the mystery and unknown. Sad to say, despite recent expeditions, there is unlikely to be a population of Brontosaurs lurking up the River Congo but there are other surprises. Mauritius, now recognised as an oceanic hot spot, was the home of the dodo (*Raphus cucullatus*), discovered in 1598 and extinct by 1670. The dodo was the epitome of the absurd, a fat, waddling, flightless pigeon permanently clad in juvenile plumage, too stupid to avoid being eaten into extinction. Analyses of fossil bone structures reveal an athletic, leggy creature, confirmed by the earliest illustrations. The dodo was an innovative design but fatally maladapted to human interference.

Continental hot spots also hold surprises. Several virulent diseases such as Lassa fever and Ebola virus seem to flare from African rainforest hot spots. Ebola outbreaks do not only kill humans. In 1994 a forty-strong chimpanzee (*Pan troglodytes*) clan in the Tai Forest of Cote d'Ivoire lost twelve to Ebola. Ebola virus replication is error prone; the frequent mutations that result permit infection of a variety of species. The Tai chimps were proficient hunters of small mammals and perhaps picked up the disease from the rodents that have boomed since 1990, when Liberian refugees fleeing civil war crossed into the area. The dodo and the Tai chimps may seem idiosyncratic but prompt important questions. How do human pressures, whether hungry sailors or refugees, impact natural systems? Chapter 4 describes species extinctions, ecosystem loss and the underlying causes for the degradation of biodiversity.

Summary

- Biodiversity is categorised as genetic, organismal and ecological. Domestic and cultural categories can be added.
- Measures of the richess of extent of any category remain fraught. Estimates of total species alive nowadays range between 10 million and 30 million.
- Functional biodiversity is very important to planetary health. Ecosystems function, as a result of species activities, providing services to the wider environment.

Discussion questions

1   Is the giant panda more important than the polio virus?
2   If all the wetlands in your country were gone, what would have been lost?
3   The biodiversity of islands can be strange, special and frightening. Why?

Further reading

See also

Why biodiversity matters, Chapter 1 pp18–19.
What do all those species do? Part I and Part II, Chapter 2 pp77–80.
Current losses of biodiversity, Chapter 4 pp127–137.

General further reading

Bibby, C.J. (ed.) (1992) *Putting Biodiversity on the Map*. International Council for Bird Preservation, Cambridge.

Degaard, F.A. (2000) How many species of Arthropods? Erwin's estimate revisited. *Biological Journal of the Linnean Society*, 71: 583–597.

Doolittle, W.F. (2000) Uprooting the tree of life. *Scientific American*, February: 72–77.

Gewin, V. (2002) All living things, online. *Nature*, 418: 362–363.

Groombridge, B. (ed.) (1992) *Global Biodiversity: Status of the Earth's Living Resources*. Chapman and Hall, London.
Essential and detailed inventory of biodiversity patterns.

Groombridge, B. and Jenkins, M.D. (2002) *World Atlas of Biodiversity: Earth's Living Resources in the 21st Century*. University of California Press, Berkeley, CA.

Hawksworth, D.L. and Kalin-Arroyo, M.T. (eds) (1995) *Magnitude and Distribution of Biodiversity*. In V.H. Heywood (ed.) *Global Biodiversity Assessment*. Cambridge University Press, Cambridge.
Essential and detailed inventory of biodiversity patterns.

Kerr, R.A. (2000) Deep life in the slow, slow lane. *Science*, 296: 1056–1058.

May, R.M. (1988) How many species are there on earth? *Science*, 241: 1441–1449.
Robert May reviews revived interest in this question prompted by rise of biodiversity research in 1980s.

Wilson, E.O. (1994) *The Diversity of Life*. Penguin, London.
Reviews extent of biodiversity, especially importance of hot spots.

Wilson, E.O. (2003) The Encyclopaedia of life. *Trends in Ecology and Evolution*, 18: 77–80.

Biodiversity hotspots

http://www.biodiversityhotspots.org/
If you want to see examples of biodiversity in the domestic/cultural realms, apparently a long way from traditional biology, try http://www.compasnet.org, which encourages and supports local development, traditional knowledge and experimentation for development.

If you would like the opportunity to have a species named after yourself and sponsor further research, go to http://www.biopat.de/

◼4 Extinction

Extinction and habitat loss epitomise our sense of a biodiversity crises. This chapter covers:

- **Extinction rates and ecosystem loss**
- **Causes of extinction**
- **Human pressures on biodiversity**
- **Valuing biodiversity**

Extinction can bring with it all the familiarity and fame akin to that of a dead pop star. The dodo and the dinosaurs are household names, evocative of failure. Neither deserves this epitaph and the confusion between types of extinction, natural versus anthropogenic, hinders our understanding of the rates and causes of loss. This chapter describes losses to biodiversity during the current crisis, driven by human activities, the causes, both ecological and economic and consequences.

Current losses of biodiversity

Prophets and loss: estimating current extinction rates

The fate of all species is extinction, a natural process unleashing evolutionary creativity. Quantitative and qualitative changes to extinction rates define the current biodiversity crisis. However, the growing awareness of extinction, fuelled by slogans such as 'extinction is forever', has proven difficult to quantify. Several approaches have been used to measure recent extinction rates and project future trends.

Confirmed extinctions

The precise start of the current crisis, defined by the impact of humans, is hard to delimit. Unusual bursts of extinction coincided with human arrival in Australia (30,000 to 50,000 years ago), North and South America (11,000 to 12,000 years ago), Madagascar (1400 years ago) and New Zealand (1000 years ago). In every case the same types of animals were disproportionately affected, with severe losses of large mammals (**megafauna**) and flightless birds. In North America the losses track the spread of early indigenous people down the continent and have been dubbed the Overkill Theory (Box 15). No such losses are evident in Africa, which may refute human impacts as a cause or may reflect the much longer co-evolution of humans and African wildlife.

The evidence of megafauna overkill by people is circumstantial but throughout the last 40,000 years, waves of extinction have hit continents and islands coincident with human

Box 15

The Clovis overkill extinctions

A wave of extinctions hit North American mammals between 11,000 and 12,000 years ago. Over the comparatively short span of 1000 to 2000 years, at least 335 species were lost, from small insectivores to largest elephants. Losses of megafauna were especially dramatic, e.g. 90 per cent of large herbivores. The losses are coincident with climatic changes but peculiar in their selectivity of large native species and progress across the continent. These extinctions have been linked to the spread of humans, invading via a land bridge from Asia and reaching what is now the southern United States by 11,000 years ago. This Clovis culture invasion may have brought in new hunting techniques, the hunters butchering their way south. The naive native megafauna were hopelessly vulnerable and extermination of these prey species led to losses amongst native carnivores. Mastodons, mammoths and giant ground sloths were wiped out. Invading mammals which spread in from Asia alongside the hunters and which were used to human pressure survived. The **Clovis overkill** remains contentious. Perhaps a combination of climatic first strike and human hunter follow-up combined to destroy the giant mammals. Alternatively disease, carried by humans or the dogs that lived with them, might have jumped across species onto the megafauna as humans spread over the continents; this is the hyperdisease hypothesis. Diseases are particularly virulent to new host species which lack in any immunity from historical adaptation, as we are finding to our cost from the spread of of HIV and Ebola fever (see Figure 4.1).

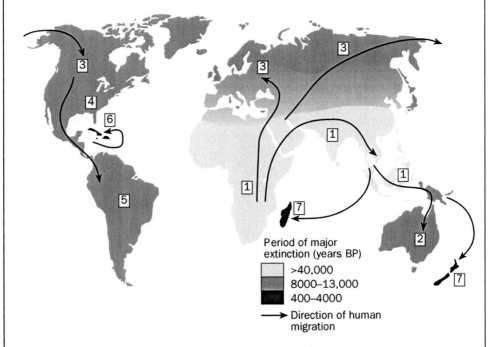

Figure 4.1 Waves of major extinctions in the wake of humanity's advance across the planet

Source: Redrawn from Martin P. S. Prehistoric overkill. In Martin P. S. and Wright, H.E. jr. (eds) (1967) *Pleistocene Extinctions*. Yale University Press, New York.

arrival. Extinctions on oceanic islands are more convincingly due to humans. Losses occurred before European colonisation: the Moa were eradicated by the Maoris in New Zealand, whilst in Hawaii between thirty-five and fifty-five birds went extinct during the Polynesian colonisation. Estimates of total bird species losses from Pacific islands are at least 2000 species, 20 per cent of all known bird species (see Figure 4.2).

The first documented extinction is given by Pliny the Elder writing in AD77. He records the loss of a plant called Silphium, probably an umbellifer. This had been widely cultivated for at least three centuries as cattle feed, a flavouring and for varied pharmaceutical uses. Pliny describes the dramatic decline in crops so that in his lifetime the last known stalk was worth its weight in gold and sent as a gift to Emperor Nero. Pliny attributes the extinction to a mix of short-term profiteering from over-grazing and bad land management by absentee landlords. The change in landownership came in the wake of failed revolt against Roman rule. Loss of Silphium was accelerated by vengeful local farmers driving flocks onto Silphium fields from which they had been excluded. Over-exploitation, civil strife, land-use changes? Pliny could be describing the present-day threat to the mountain gorilla or tigers.

Documented extinctions have typically been recorded from 1600, following in the wake of the European global exploration. Recorded extinctions are taxonomically biased to mammals and birds, with molluscs an unexpected third category due to intense work on some island species and geographically biased towards Europe, North America and island

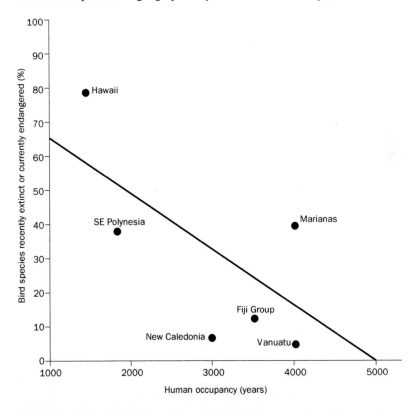

Figure 4.2 Birds recently extinct and endangered on Pacific islands relative to the length of human occupation. The numbers of species in immediate danger are highest on islands most recently occupied. Species vulnerable to human pressures ('wimp species') have long been lost from islands occupied earlier. The increased threat to the Marianas bird fauna is due to recent colonisation of the islands by the brown tree snake

Source: Redrawn from Lawton and May (1995).

Table 4.1 *Known extinctions up to 1995*

	Total	On islands
Mammals	58	34
Birds	115	104
Molluscs	191	151
Other animals	120	74
Higher plants	517	196

Source: Groombridge 1992

systems, though the latter reflects a genuine vulnerability of island faunas (Table 4.1). Some supposedly extinct species cling on and have been rediscovered, so-called **Lazarus species**, e.g. the Mount Kupe bush shrike featured later in this chapter. Some extant species number so few individuals they are doomed to extinction, dubbed **living dead species**, such as the poor *Partula turgida* of Box 16.

Whilst losses of mammals, birds and molluscs have been fairly well documented, scarcely sixty insect species are known to have become extinct and of these half are Hawaiian examples.

Predicting extinction rates

Estimates of extinction rates rely on extrapolating data for known patterns of loss, much as estimates of total diversity rely on expansions of models. Five approaches have been used:

- estimates from recent past extinction rates
- predictions from habitat loss
- predictions from the changing status of threatened taxa
- molecular
- energy use.

Estimates from recent past extinction rates

Extinction rates from fossil records are best documented for marine molluscs, whilst birds and mammals have a very patchy fossil record. Fossil data suggest mammal extinctions at a rate of one species every two hundred years, compared to known losses of at least twenty species during the twentieth century. For birds extinction rates in the last two thousand years are a thousandfold increase on loss rates prior to major human impacts (Figure 4.3). Estimates are also biased by the number of recorded extinctions from oceanic islands in the Pacific. For birds and mammals combined, a general estimate of extinction rates is a 1 per cent loss during the twentieth century.

Predictions from habitat loss

Ecological science provides a host of examples of species–area relationships. Essentially the larger the area of habitat, the more species it supports. As a gross rule of thumb, a 90 per cent decrease in area results in a 50 per cent loss of species. One of the main causes of extinction is habitat loss. As the area of habitat declines, species–area relationships predict that species numbers will decline. For habitats where loss rates have been measured and species–area patterns are known, extinctions can be predicted. Estimates based on habitat loss for rainforest species vary with assumptions of forest loss rates, species richness in proportion to area, and variations between precise type of forest used for original data and species-area model used. Estimates of losses up till 2020 vary between 2 per cent and 25 per cent, depending on taxa. Estimates of forest loss give a general prediction of extinction of between 1–10 per cent of all the world's species up till 2020, a thousandfold or ten thousandfold increase on background extinction rates (see Figures 4.4 and 4.5).

Box 16

A documented extinction: *Partula turgida*, a snail

At 17.30 hours on 1 January 1996 the last individual of the snail *Partula turgida* from French Polynesia died in London Zoo (Plate 16). To time an extinction so precisely is very unusual. The snail was rescued from the wild in 1991 as part of a conservation programme for the endemic snails of Pacific islands set up in 1987, now nurturing thirty-three taxa dispersed between eighteen captive stock refuges. Many Pacific island endemic snails have been driven to actual or imminent extinction by release of a predatory snail *Euglandinia rosea*. This fast moving mollusc was itself introduced as a **biocontrol** for giant African land snails (*Achatina fulica*). African snails were imported as food, but escaped and flourished, becoming a pest species attacking crops. *Partula* species have aroused attention not simply because of documented losses, detectable from remnant shells, but also from their fame as examples of evolutionary diversification, with some species apparently limited to individual valleys on some islands. In 1994 three species of *Partula* were reintroduced into a small enclosure in Moorea from captive breeding programmes in the United Kingdom. The walled enclosure was further defended by electric fencing. Unfortunately monitoring in 1995 showed *Euglandinia* had got in using fallen vegetation to bridge the barriers and had killed all the snails. More encouragingly, shells of young snails showed that the *Partula* were capable of breeding following release. The work continues.

Plate 16 *Partula turgida*. The last individual of this species before its extinction in 1996. London Zoo is home to captive populations of several *Partula* species, rescued from imminent extinction in the wild and now breeding in captivity. Their offspring are the foundation of reintroduction programmes.

Predictions from the changing status of threatened taxa

The International Union for the Conservation of Nature maintains internationally recognised lists of threatened taxa, the **Red Lists** and **Red Data Books**. These categorise taxa according to the intensity of the threat and include acknowledged extinct species. The rates at which species are added to these lists or move between categories can be used to estimate extinction rates.

Figure 4.3 The Dodo. Apocryphal symbol of extinction as doomed, dumpy convenience food but recently revealed as a lithe ground dweller, probably well suited to its island habitat but not to the arrival of humans

Source: Redrawn from *New Scientist*, 28 August 1993.

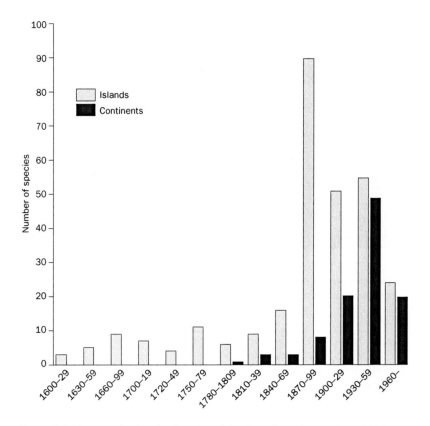

Figure 4.4 Known animal extinctions from islands and continents since 1600

Source: Redrawn from Heywood (1995).

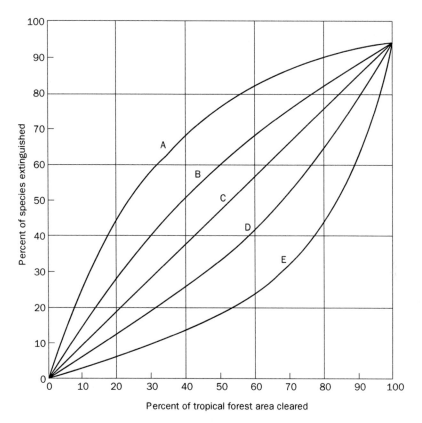

Figure 4.5 Different models of species loss with forest clearance discussed by Lovejoy in his seminal 1980 paper

Source: Redrawn from Lovejoy (1980).

Between 1986 and 1990 the animal species list has increased by more than 30 per cent but only 15 vertebrates and 33 animals in total have been added to the extinct category, while 163 plant species have been added to the lists. For most taxa the lack of data for most species makes this approach poor, but for birds and mammals gives predictions of loss of half the extant species within 200 to 300 years, a time to extinction for half the extant vertebrate species of 7000 years, and 3000 years to wipe out half the known plant species.

Estimates from these techniques coincide with those derived from species–area estimates for bird and mammal losses. There is general agreement. For birds, mammals and perhaps other terrestrial vertebrates, current estimates of loss rates predict 50 per cent of species alive now will be lost within 200 to 500 years.

Molecular

Phylogenetic relationships can be explored at the molecular level, though the technology for detailed and extensive investigations has only recently been available. Molecular data from present-day species gives clues to the past, present and maybe future of a taxon. The story of a taxon's evolutionary history can be unravelled, including estimates of extinction rates that find an echo in the molecular baggage. Molecular phylogenies define individual lineages within larger taxa and the times of splits.

Analysis of phylogenetic trees defined by such precise molecular information has shown that extinction rates can be estimated for some taxa by comparing observed data to predictions of lineage history based on a model that supposes the probability of a new lineage starting or ending (extinction) is equal and constant throughout the taxon's history.

Energy use

Extinctions can be correlated with measures of human energy consumption, a surrogate measure for the pressure of human population growth and activity. Known species extinctions can be measured against known increases in energy consumption and future extinctions predicted by extrapolation against estimates of future energy consumption. One general estimate using this technique, based on energy increase predicted by the Brundtland Report in 1987, suggests that the time to extinction for half the remaining bird and mammal species is only thirty years. Comparison of known species extinctions of taxa such as butterflies from countries with reliable measures of energy use in the twentieth century do not match the predicted loss rates from energy models, but energy-based models may correlate well with losses of scattered populations, reducing ranges and total populations.

Haemorrhaging biodiversity

The advent of the biodiversity agenda focused attempts to measure extinction rates globally and locally. WWF and UNEP attempted a synthesis of loss rates between 1970 and 2000 for 694 populations of vertebrates which were sufficiently well studied to provide robust data. The resulting Living Planet Index suggested that most populations were declining, with marked variations between habitats. Mean changes from 1970 to 2000 for forest dwelling species were –15 per cent, for freshwater species –35 per cent and for marine species –54 per cent. Across the four biomes for which there were sufficient data (tropical forest, boreal forest, sea-grass and mangroves) vertebrate populations were changing by –0.7 per cent to –1.1 per cent annually. There are problems with these measures. Most taxa and biomes are missed out, there are variations in data collection and the results do not highlight losses of keystone species which play a vital role in driving ecosystem functions. However, there are benefits. The data generated headlines proclaiming we are losing between 0.5 and 1.0 per cent of wild nature each year, a slightly abstract figure that chimed with the rapid losses of many familiar species. For example in the United Kingdom, the house sparrow (*Passer domesticus*) has undergone a 50–60 per cent population decline from 1970, provoking popular media coverage. The WWF and UNEP analysis also put financial estimates to the losses, putting the annual loss of goods and services due to population and habitat declines at US$250 million a year. Even if species are not yet extinct, their decline is damaging biodiversity services. Extinction is the final curtain but we need to know how loss of species is degrading services and sustain species and habitats as functionally viable populations and areas.

Where detailed data are available, alarming results have been produced. In 2004 a synthesis of data on declines and extinctions amongst the very well known and thoroughly recorded British butterflies, bird and flowering plants was published. Changes to range (a good surrogate measure of population abundance) were analysed for the late twentieth century. Native plants had declined by 28 per cent in forty years, birds by 54 per cent in twenty years and butterflies by 71 per cent in twenty years. The loss of butterflies emphasised the problems of fragmentation and local loss of populations. Some species may still be fairly widespread in the United Kingdom but what were once continuous ranges are now punched full of gaps and essentially falling apart.

These analyses have highlighted the need to measure the changes to the range of species and habitats. The emphasis on final extinction is too late. Most extinctions go unseen; we rely on estimates based on a few taxa (e.g. birds), which may not be representative. Some species may already be doomed even though there are remnant populations whiling away the days to oblivion though the fatal damage has already been done, a process for which the unfortunate ecological term is relaxation. A better way of describing this is extinction debt. One estimate for African forest primates is that 30 per cent of species fall into this fauna of the living dead. Estimates of population changes for mammals show the extent of population contractions. An analysis of 173 declining mammals, compared against their ranges in the nineteenth century, shows declines throughout the world, for example Africa, 52 species, mean range contraction 72 per cent, and South America, 17 species, mean range contraction 15 per cent.

Global-scale estimates of extinction have used the predicted impacts of climate change and the resulting changes of habitat area to estimate likely losses. Combining data for a range of biomes (e.g. boreal forest, grassland) and taxa about which we know enough to make robust estimates (e.g. mammals, butterflies) the results suggested that by 2050, based on mid-range climate change projections, between 15 and 37 per cent of species would be 'committed to extinction', even if it took much longer for the final populations to snuff out. Such estimates inevitably cause debate, not least because of clumsy reports in the press misinterpreting the findings as a million or more species would definitely be extinct by 2050. Nonetheless the figures make bleak reading.

Ecosystem loss rates

Loss of habitats has become a touchstone for conservationists. 'An area equivalent to the size of Wales' has become almost an SI unit for measurement of habitat destruction. Habitat loss deserves attention as both a measure of biodiversity loss and as a primary cause of species extinction. However, ecosystems are hard to define and dynamic; their precise size, boundaries, time scale of change and threshold beyond which they have been irredeemably wrecked are imprecise. Habitats can alter from one type to another; whilst one species is lost, another may be able to establish.

There are few reliable measures of global ecosystem loss rates other than for tropical rainforests, which have attracted special attention because these forests harbour the greatest species diversity. Estimates vary with precise definition of forest types (e.g. closed versus open) and as better coverage of more countries has been added to the database. Global percentage losses have been estimated as 1976–80 0.6 per cent, 1981–85 0.62 per cent and 1981–90 0.9 per cent and 1990–2000 0.8 per cent. Losses vary between countries. High losses, between 1981 and 1988, include Costa Rica (4.0–7.6 per cent), Thailand (2.4–2.5 per cent) and Brazil (0.4–2.2 per cent). Losses are not solely a phenomenon of recent decades. Estimates of forest loss in Africa across recent centuries are 96,000 to 226,000 sq. km before 1650, 470,000 sq. km in the mid-nineteenth century up till 1978, and 51,000 sq. km from 1981 to 1990.

Udvardy's global ecosystem classification (Chapter 3) has been used as a basis to assess loss and degradation of ecosystems on a global scale. Major vegetation types, biomes, were then assessed within each. The biomes were identified as provinces reflecting regional variations in form. Table 4.2 lists Afrotropical and Neotropical examples, to compare with Table 3.9.

Defining the endangered

Animals sporting such labels as 'endangered' or 'rare' are the familiars of umpteen television documentaries and public campaigns, yet these terms can be bandied about with no precise definition. The California condor (*Gymnogyps californianus*) is rare; there

Table 4.2 Loss rates of example ecosystems, based on Udvardy's global classification. The percentage totally undisturbed is given, along with an index based on per cent partially disturbed or dominated by humans

Realm	Biome and province	Area, km²	% totally undisturbed	Index of damage (0 = total loss, 100 = intact)
Afrotropical	**Tropical Humid Forest**			
	Congo rainforest	2,195,019	61.2	66.6
	Guinean rainforest	709,112	8.4	13.3
	Malagasy rainforest	147,862	39.4	39.4
	Evergreen Sclerophyllous vegetation			
	Cape Sclerophyll	99,663	13.5	18.5
Neotropical	**Tropical Humid Forest**			
	Amazonian	2,864,623	98	98
	Campechean	279,695	33.6	35.7
	Colombian coastal	273,266	45.9	52.3
	Guyanan	1,090,396	94.5	95.1
	Mudieran	1,988,840	81.5	82
	Panamanian	128,872	90	90
	Atlantic forest	223,944	6.5	12.6
	Evergreen Sclerophyllous vegetation			
	Chilean sclerophyll	47,988	36.1	45.8

Source: Hannah *et al.* 1995

are only a few individuals alive. Cuvier's beaked whale (*Ziphus cavirostris*) is rare, only a few having been seen off the Pacific and Atlantic coasts of Canada but there are more, perhaps many more, out there in the wild. The scarce emerald damselfly (*Lestes dryas*) is rare in the United Kingdom; it was thought extinct in the 1970s, rediscovered at a few sites in the 1980s, but widespread across continental Europe and Asia. Precise definitions are not merely the realm of experts but important for practical purposes, to raise awareness, to define the current threat to a species survival. Changes in status can be monitored and priorities for action decided.

Reviewing the concept of rarity itself, Gaston (1994) showed that definitions varied with taxa, measures of abundance and extent of range. Population abundance or spatial range could be used quite independently. He concludes that many studies classify a certain number of species as rare because that is the number which feels right and Gaston suggested a practical threshold for classification as rare as the 25 per cent of species in a taxon with the lowest abundance or range size. More abundant species can be described as common, those with a wider range as widespread. Even so, stating the scale at which judgements of rarity are made is an important part of any categorisation.

The term 'threatened' is widely recognised as an inclusive label for different degrees of danger to species' survival. In the 1960s the IUCN started to compile the Red Data Books, cataloguing known threatened species from around the world. The books have now spawned the Red Lists as the quantity of data grows, updated every two years by the IUCN and the **World Conservation Monitoring Centre (WCMC)** using advice from specialist groups. Box 17 outlines the IUCN categories.

Box 17

Red Data categories

In 1994 a revised ten-category classification system replaced the previous six-category system of extinct, endangered, vulnerable, rare, indeterminate and insufficiently known. The new categories are as follows:

Extinct No reasonable doubt that the last individual has died.

Extinct in wild Only known to survive in captivity or naturalised well outside past range.

Critically endangered Extremely high risk of extinction in the wild in the immediate future.

Endangered High risk of extinction in wild in the near future.

Vulnerable High risk of extinction in wild in the medium-term future.

Conservation dependent Taxon dependent on conservation programme which if stopped would place taxon into one of the above categories within five years.

Near threatened Taxon close to qualifying for one of the above categories.

Least concern Taxon does not qualify for any of the above criteria.

Data deficient Data insufficient to categorise taxon but listing highlights requirement for research, perhaps acknowledging suspicion that taxon warrants classification.

Not evaluated Taxon not assessed.

Problems inevitably arise with such classifications.

- *Bias:* the categories contain only known species. The undiscovered ones plus species found but not yet described are omitted. The great majority of species are therefore excluded and the published lists could divert attention from this unknown majority. Coverage for better known taxa varies: birds nearly 100 per cent, mammals 50 per cent, reptiles 20 per cent, amphibians 10 per cent, fish 5 per cent.
- *Species based:* the Red Data Books deal with individual species. There is no attempt to assess the threat to higher taxonomic levels, which may represent more fundamental differences and variety or specialness.
- *Lack of objectivity:* the superficially neat, objective classification is based on expert opinion, so is subjective. Risk is poorly quantified, such as the percentage risk of extinction over a defined time period.
- *Some criteria are not linked to threat:* Red Data classification can be based on protection accorded to a species or unhelpful criteria, e.g. 'insufficiently known' (which applies to most life on the planet). Other potentially useful data such as rate of population decline are not routinely used.

The causes of extinction

We are all versed in the many causes of extinction, whether it is chopping down the Amazon rainforests to open up farmland, hunting tigers so that their bones can be ground into cold cures or beach front discotheques disorientating hatchling turtles. The individual cases provide such a litany of woe that generally applicable patterns and processes are difficult to decipher.

The ecological causes of extinctions are natural processes that put wildlife at risk. These comprise ultimate causes that drive populations into decline or keep them rare, and the proximate causes, those final straws that snuff out the last of a line. The pressures of human activities are direct, proximate threats, for example exploitation, extermination, habitat destruction, introduced species, pollution and ecosystem breakdown. These threats are driven by four main characteristics of human systems: resource pressures, cultural attitudes, institutional policies and the failure of economics to value biodiversity properly.

Ecological causes of extinction

Natural processes can put species at risk from extinction. These ecological processes are divided into two: **ultimate causes** that drive decline and rarity and **proximate causes** that wipe out remnant populations.

Ultimate factors

Population abundance and geographic range

As a general rule species with a wide range have higher populations. This results in an ominous double jeopardy. Reduce a population and its range will contract. Diminish the range and the population will fall. Why abundance and range are correlated is unclear. An intuitively attractive possibility is that species with broad niches can be widespread and locally abundant, able to exploit diverse resources. Abundance and range can also be analysed using **metapopulation** models. A metapopulation is a set of separate populations linked by emigration and immigration. Models of metapopulation dynamics support the observation that wider ranges (more patches occupied) and higher populations (within patches) are linked, because of **rescue effects** (arrivals top up populations in patches near extinction) and higher populations compensating for losses during migration.

Worryingly there is a marked trend to smaller ranges and population sizes towards the tropics. Tropical species, the heart of Earth's biodiversity, may be naturally more vulnerable to extinction.

Patchy distributions within overall range

Populations can be unevenly distributed across the total range; sometimes the differences are so extreme that high density areas act as **sources** topping up numbers in less favourable (**sink**) sites. A population in decline, contracting in range, may fragment. Isolated, smaller populations are more vulnerable to extinction because they are prone to local catastrophes and too far for rescue immigration or recolonisation. Fragments left behind in peripheral sites, beyond immigration range from core sources, may be doomed even without additional human pressures.

One large population is safer if demographic stochasticity is the main proximate threat to survival. Many, albeit smaller, populations are safer from the danger of large, catastrophic disasters such as fire and disease. Some patches may escape. This security decreases if

Plate 17 Habitat fragmentation. This South American tropical forest is a 1 kilometre square remnant, the last home of the three toed jacamar. The fragment may be too small to support a viable population, so the individuals are living dead, the species doomed once the survivors die. Hence the presence of the desperate bird watchers.

events in individual patches are spatially synchronous. As with others of these ultimate causes, evidence is sparse and contradictory. In general, patchy populations show synchrony even if patches are hundreds of kilometres apart, so protection by this means may be rare. Given that larger populations are also less vulnerable to genetic failure and social breakdown, what little evidence there is suggests a few large populations are better than many smaller ones (see Plate 17).

Body size and trophic position

Evidence from animals suggests that big is bad. The megafauna overkills are an extreme example, with human attention focused on large species that provided more meat and that were a direct threat to people. Larger animals are often higher in food chains, so are vulnerable to any disruption to trophic levels lower down, whether losses of lower levels or damaging factors concentrating in higher levels such as the famous pesticide impacts.

Both these apparently simple inferences are very unreliable. Body size alone is a poor predictor of immediate vulnerability to extinction. Instead body, population and range size variously combine to give different patterns. Smaller species have contradictory attributes. Many show more frequent and larger population fluctuations (an increased risk factor) but tend to breed more rapidly and abundantly (decreasing risk). Part of the confusion arises because body size correlates with so many other ecological attributes of a species.

Colonisation ability

Colonisation, a process combining both dispersal to new sites and successful establishment, limits the ranges of many species, both common and rare. Colonisation ability becomes an important factor if rare species are generally less effective colonists than the common ones. Different species with good powers of dispersal have ranges of many different sizes. Poor dispersers show less variation, possessing small to medium range sizes. Given the links between range size and abundance, poor dispersers may be at increased risk. Conversely some taxa have lost their powers of dispersal where emigration is almost inevitably a doomed enterprise.

Rarity shows a weak link with establishment ability, but evidence is confounded by other constraints, in particular those which restrict the ability for rapid reproduction, which is generally useful for establishment.

Historical echoes

Species may be rare due to historical events of which we have little idea. In addition some lineages are more or less susceptible to extinctions, though why is unclear. The vulnerable lines are those with small ranges and low abundance.

In conclusion the ultimate ecological causes of rarity appear idiosyncratic with cause and effect difficult to distinguish and patterns and process varying across time and spatial scales. Combining the lessons of historic extinctions and recent losses, Raup (1993) picks out five generalities.

- Species with small populations are more vulnerable.
- Species with limited ranges are more vulnerable.
- Extinction of widespread species is increased following an environmental first strike.
- Extinction of widespread species is favoured by stresses outside their normal range of experience; the return time scale of the threat is important.
- Simultaneous extinction of many species requires stresses that cut across ecological lines.

Proximate factors

Once a population is small or fragmented, extinction can be caused by many processes.

Demographic stochasticity

Patterns of reproduction and survival are the demography of a population. Small populations (from tens to hundreds of individuals) are vulnerable to random (stochastic) misfortunes such as being unable to find a mate or suffering losses to predation. Such events would be insignificant to the overall survival of a larger population but may finish off small numbers. The helmeted honeyeater (*Lichenostomus melaops cassidix*), a rare bird of Victoria State, Australia, is threatened by such accidents. The honeyeater is endemic to riverside forest, its range contracting in the face of human habitat clearance. The last ten colonies collapsed down to just one containing sixty birds. There is now an appreciable risk of extinction due to natural vagaries of mating success and nesting. In particular, loss of a male during breeding will result in death of the pair's brood even if the female is still alive (see Figure 4.6).

Environmental stochasticity

Random fluctuations to a species' habitat, e.g. variations in weather or food supply, represent a danger to small populations. Again such changes would be little threat to the common or widespread, which absorb local losses, but isolated pockets of a rare taxa can be lost with no opportunity to recolonise.

UK research into the survival of pond fauna has shown that occasional drying out of a pond causes marked losses of permanent water species. Temporary pond species may colonise, their diversity often lower. However, a survey of temporary and permanent ponds in Oxfordshire revealed that when ponds were ranked by rarity of species rather than diversity, four of the top five ponds, out of thirty-nine surveyed, were temporary. One of

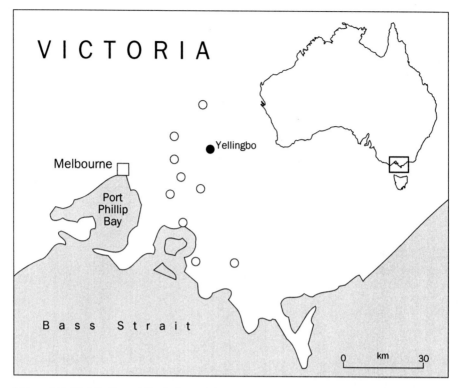

Figure 4.6 Distribution of the helmeted honeyeater in Australia. Only one colony (solid dot) remains. Open circles indicate old sites, now extinct. The bird is vulnerable to extinction from simple accidents, with all its eggs in one site basket

Source: Redrawn from MacCarthy *et al.* (1994).

the temporary ponds was home to water beetle, *Haliplus furcata*, classified in UK lists as in the most critical Red Data Book category 'RDB1, endangered'.

Genetic failure

Small populations risk genetic failure. Inbreeding loses genetic variability. Offspring become increasingly **homozygous** (i.e. similar genetic make-up) lowering short-term fitness such as resistance to disease, and longer-term genetic variation necessary to survive evolutionary pressures. Inbreeding may result from new, usually harmful, mutations or established, disadvantageous characteristics finding expression as more individuals carry them. The process can snowball as more parents carry disadvantageous genes, lowering the chance that one parent may provide a dominant, compensating gene to offspring.

A study of forty adders (*Vipera beras*), a fragment population in southern Sweden, isolated in a 1000 by 20–500 m grassland some 20 km from the nearest populations, is an unusual example of evidence of inbreeding in the wild. Over the seven-year study period only between one and fifteen males and between two and eighteen females bred in any one year. Smaller litter sizes, increased deformity, stillbirth and marked genetic uniformity were characteristic of the population compared to adders from extensive populations. This degradation had occurred within the space of ten years since the population was cut off.

Natural catastrophes

Natural catastrophes could be regarded as extreme examples from a continuum of increasingly severe environmental stochasticity but they do represent a qualitative step up, including the likes of volcanic eruptions, fire and flood. Volcanoes are explosively destructive, fires are not simply hot but burn, and floods are a physically destructive force.

Natural disasters have been a factor in the decline of the Samoan and Tongan fruit bats (*Pteropus samoensis* and *P. tonganus*) in American Samoa. Numbers in the mid-1980s were estimated at 1500 and 12,000 respectively. Hurricanes in 1990 and 1991 caused 80–90 per cent mortality, killing bats and wrecking forest habitat. Losses were worsened as human hunters took an added toll. The bats had always been harvested in small numbers but the hurricanes increased accessibility of roosts and many bats moved into village plantations. The post-hurricane harvest of 3400 actually exceeded first estimates of the entire surviving bat populations. Bat numbers are now about 400 and 2500 respectively (see Figure 4.7).

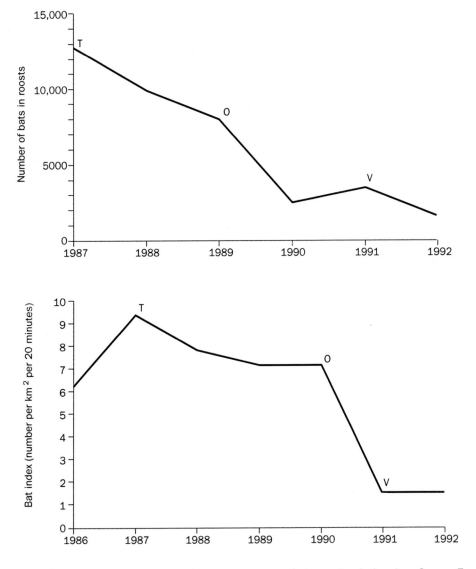

Figure 4.7 Decline of bats, measured at roosts or counted when active, in American Samoa. T, O and V indicate different hurricane events. Bat abundance declined severely due to natural disasters

Source: Redrawn from Craig *et al.* (1994).

Extinction: human pressures and economic failures

The current threat to biodiversity, perhaps equivalent to a sixth major mass extinction, is caused by the impact of humans. Our actions can cause the decline and extinction of species in very direct ways, the **proximate causes** of extinction. However, these individual cases are driven by the four underlying **ultimate causes** of resource use, cultural attitudes, institutional failure and perhaps most pressing of all, the failure of existing economic systems to fully capture the value of biodiversity.

Human impacts on biodiversity: proximate causes

Exploitation
Exploitation is the direct use of a species, thus driving down numbers. The famous examples, such as poaching of rhinos for their horns, sometimes eclipse the massive global trade in many species. Examples include the expanding Asian trade in terrapin and tortoise products for medicine and diet supplements. Trade has been cited as a major cause of the catastrophic decline of species in southern China; the Chinese markets have now expanded to exploit stocks from other Asian countries. Exploitation of at least six species is reported at unsustainable levels including trade in some species that is illegal under the CITES treaty.

Habitat destruction
Habitat destruction is typically the result of exploitation or conversion of wild habitat. A typical case causing problems is the destruction of parts of the Khao Sam Roi Yot National Park in Thailand, a globally important area of coastal wetland. The area had been used historically by local people harvesting wild prawn. Government and industry have promoted intensive prawn farms, using artificially managed lagoons. Land prices rose and farmers from outside claimed ownership by using dubious documentation. Hotels and golf courses now occupy additional habitat. This destruction of primary habitat, combining conversion to farmland, coercive and aggressive acquisition and weak protection, is typical of many developing countries.

Introductions and exterminations
Humans have caused losses by the intentional extermination of species and introduction of others. In 1995 African wild dogs in Namibia, numbered at 250, out of a total population in Africa of 5000, were hunted by farmers following a pack kill of a calf; 32 dogs were killed. Wild dogs were still classified as vermin in Namibia and could be legally killed. By contrast the wild dog is one of only five mammals listed as Specially Protected Animals in Zimbabwe's Parks and Wildlife Act.

Extinctions caused by introduced species have famously devastated many oceanic island systems. A more insidious problem, extinction through genetic dilution and blurring with introductions, has come to the fore in recent years. The world's most endangered dog, the Ethiopian wolf (*Canis simensis*), is vulnerable to this fate. Reduced to fewer than 500 individuals by disease such as rabies and canine distemper, both perhaps spread by domestic dogs, hybridisation with dogs is further eroding the species.

Pollution
Pollution may be one of the earliest forms of damage which raised public awareness of environmental degradation but remains just as widespread. Much of the South American rainforest is currently threatened by pollution of waterways from toxic waste products of

gold mining, notably mercury and cyanide. Guyana suffered one of the worst mining disasters in history in 1995, all the nastier given its predictability. Cyanide-rich slurry, including arsenic and copper, breached slurry tanks, spilling 3 million cubic metres into the River Essequibo. Guyana's president declared an environmental disaster zone, having previously welcomed the large-scale mining as turning 'our mudland into the gold land of the future'. Goldstar Resources Limited, the American mining company involved, explained its operations in Guyana more honestly. Goldstar had 'looked specifically at the Guyana Shield because of increased pressure from environmentalists and government in the USA'.

Ecosystem cascades

Human activities which, even if not directly destructive, spark effects that ripple through ecosystems as natural interactions are disrupted have become a common threat (see Box 18). A globally commonplace cascade impact is due to overenrichment of nutrients in seas and lakes. The abundant supply can cause a switch from a diverse community to one dominated by a few species able to thrive on the enrichment. The destructive imbalance snowballs as more and more diversity is lost. This overenrichment is called eutrophication and has been linked to red tides of massive blooms of algae. Bloom species often release toxic chemicals or cause further destruction when their death and decay deoxygenate the water. Recent deaths of the American manatee around Florida (over 150 out of populations of 3000) have been linked to a toxic red tide.

Box 18

The plight of Asia's vultures

Species can switch from being commonplace to threatened with alarming speed. The collapse of Asian vulture populations over the turn of the millennium is a startling example. Four species of the Genus *Gyps* occur throughout India, Pakistan and Nepal. In the 1980s they were familiar birds, often found in big cities, and one species, the white-rumped vulture (*G. begalensis*), was regarded as the most abundant large bird of prey in the world. In the late 1990s populations of three species, *G. bengalensis* the slender billed vulture, (*G. tenuirostris*) and the Indian vulture (*G. indicus*) crashed to a point where all are now regarded as critically endangered. These often unloved birds could replace icons of extinction: populations fell faster than that of the dodo and they are now more vulnerable than the tiger. Early reports suggested a disease due to inflammation of the gut. By 2003 tests had shown high residues of a painkiller drug, diclofenac, in the kidneys of birds which had died with the characteristic symptoms, but not in birds killed in other ways. This drug has only recently been introduced in veterinary practice but its use is widespread. Further work in 2004 confirmed that the drug caused fatal kidney failure at dose levels one-tenth below those known to affect mammals. The loss of vultures was not just a simple loss of biodiversity for its own sake. Feral dog populations, unaffected by the drug, have increased because there are fewer vultures to compete for carcasses, bringing with them rabies. Local customs have been affected too. Celestial burials, where human corpses are left out to be scavenged by vultures, have had to be replaced by cremation.

Mismanagement and confusion

Improvements in conservation policy and practice can be set back, reducing hoped-for protection. In 1996 the World Bank Reconstruction and Development Division pursued plans for a £1 billion dam, Nam Theun II, in Laos. The **Global Environment Facility (GEF)**, also managed by the World Bank, was working on a £9 million conservation package to protect biodiversity hot spots in Laos that include endemics such as the large antlered muntjac, discovered only in 1993. The GEF reported that the dam would cause significant environmental degradation and dropped funding, unwilling to participate in mitigation work whilst the main damage was still allowed to proceed.

Human impacts on biodiversity: the underlying human pressures on biodiversity

The damage wrought to nature by humans is seldom driven by destructive spite (see Box 19). Our actions are the result of deeper forces and failures. To understand biodiversity loss, it is vital to understand the economic and social pressures, including resource pressures, cultural problems, institutional failures and economic failures.

Box 19

The fatal lure of the Grey Wolf

The intentional extermination of a species is rare, other than attempts at pest control. There are, however, some species which face wanton persecution and seem to provoke a irresistible desire to hunt them. The Grey Wolf (*Canis lupus*) in Europe and North America is a salutary example. In these regions the Grey Wolf was hunted as vermin and for sport, leaving fragmented populations in remote wilderness. In Eurasia and North America wolf populations were at their lowest between 1930 and 1960. Research, legislation and growing public interest led to recoveries. In Europe remnant populations in Italy, Spain and Poland have grown, with migration re-establishing packs in Finland, Norway, Sweden, France, Germany and Switzerland. There is a European Wolf Network promoting recovery and there are, even (optimistic) plans to reintroduce the wolf to the United Kingdom.

However, the recovery has led to conflict. In part this is direct economic loss, for example in Spain during the 1990s this was estimated at US$1 million a year. Compensation schemes sometimes provide an answer. Public attitudes have been contradictory, sometimes turning anti-wolf where wolves have killed pets, sometimes hindering management schemes by opposing compensation payments to farmers. If this were not complicated enough, wolves are still persecuted by hunters. In Spain, Switzerland and Norway hunters have targeted wolves even though hunting is illegal, openly boasting of their kills and vowing to eradicate the wolf. There is something about the wolf that drives us to either shoot it or print it on a T-shirt.

Resource pressures

Human population growth

The global human population continues to grow rapidly. Estimated at over 5.6 billion in the mid-1990s, the total is reckoned to double by 2052. The developing world's youthful population will maintain the momentum, adding 1 billion each decade. Increased population has a direct impact by consuming more resources unsustainably and limiting regeneration (see Box 20). Indirect effects include poverty, migration and social breakdown. Population growth causes habitat loss and uses up ever more natural production. Growth will be concentrated in the developing world, in countries most vulnerable to climate change but harbouring the greatest biodiversity. Put simply humans consume 40 per cent of terrestrial primary productivity. Energy use is disproportionately high in the developed world. All other living things conform to a neat relationship between body size and energy use, but in the western world humans break this rule, consuming energy roughly equivalent to what a 30 foot high ape would use.

Box 20

Human impacts: greed and need – the bushmeat crisis

Bushmeat is any animal hunted from the wild and used for food, although the term is primarily used to mean animals taken by hunters from tropical forests. Bushmeat hunting can drive species to extinction locally but is a vital source of food and income for some of the poorest people in the world.

In the 1990s initial studies of local bushmeat use hinted at the toll taken on local wildlife. A study of two town markets in Equatorial Guinea revealed the sheer numbers of animals killed. Over 18,000 carcasses were brought in for sale over 10 months. The tally concentrated on a few species, 9 antelope, 16 primates and 3 rodents, with evidence of preference for size, neither too small nor too large. Local taboos against eating some species (e.g. Colobus monkey) were breaking down as hunters ranged further afield into new areas. The bushmeat trade follows developments such as oil and logging: these industries provoke unsustainable use and local extinctions of target species. Individual studies are now supported by wider scale estimates of the impact of bushmeat use. In the Congo basin 60 per cent of the 51 bushmeat target species were being exploited unsustainably, with the take about 2.4 times higher than the animals' productivity. Estimates of bushmeat use in Africa go as high as 5 million tons a year. However, other work in the Congo shows that for many households, bushmeat is a major source of income. Most bushmeat caught was sold at market, a more important source of income than crops or fish, especially during the lean season. Bushmeat provides both income and food security. Even if use of bushmeat could be banned, this would hit some of the poorest people in the world. Because charismatic species such as great apes are on the menu, wildlife campaigners have been drawn into the fray. As one website puts it: 'A ragged far-flung army of a few thousand commercial bushmeat hunters supported by the timber industry infrastructure will illegally shoot and butcher more than two billion dollars worth of wildlife this year, including as many as 8,000 endangered great apes'.

The problem increasingly reaches the developed world. There is evidence that eating apes is a route for diseases to jump to humans. Simian Immune Deficiency Virus may have jumped across to humans and become HIV. Antibodies to other ape viruses have been found in humans using ape bushmeat. In December 2003 the first person was imprisoned in the United Kingdom for selling bushmeat unfit for human consumption.

Consumption is growing faster than population. Examples of increased consumption, between 1950 and 1990, include fish catches up by 44 per cent, fertiliser use up 970 per cent and natural gas production up 1150 per cent. Meanwhile more of the world's populations become consumers. Increasing urbanisation degrades biodiversity by habitat loss, fragmentation, introduction of exotic species, pollution, drain of resources, disruption of natural geochemical cycles and conversion of adjacent land for farming or suburbs as some try to escape the crowded cities. In the mid 1990s 45 per cent of the world's human population lived in cities. This is predicted to grow to over 60 per cent by 2030. The effect is worsened by the growth of particularly large cities. There were 190 cities with 1 million plus inhabitants in 1975. In 2005 there are 314.

Population movements in response to economic need add to the damage. Cultural and institutional restraints collapse amongst mobile communities as they move into undeveloped land. Settlers often lack any sense of a link to the land which fosters a sense of responsibility for resources.

Drive to globalisation

The diversity of crops, agricultural techniques and production systems around the world has been replaced by an increasing reliance on a small number of crop species. These can be traded in a global market. The globalisation of economies has broken the links between the management and consumption of resources. Many people are now effectively **biosphere people** and are able to escape problems of overexploitation by shifting to a new market for resources. Individual countries that adjust economic policies to penalise destructive, exploitative trade by their own industries open themselves up to undercutting by competition from countries that do not. Free trade can undermine local markets, disrupting local social and cultural traditions which use resources sustainably. Global trade can be a direct threat to valuable species. Illegal trade is still rife. When a single Russian Siberian tiger can be worth £40,000, with the poacher paid £660, enough to support a family for two years, and its bones, organs and skins traded throughout Asia, then the global market is an irresistible threat.

Cultural problems

Cultural problems include inequality of ownership and property rights and cultural attitudes.

Inequality of ownership and property rights

Iniquitous ownership, control and trade work in favour of degradation and extinction. Globally and locally a minority owns most resources and reaps the benefits of exploitation whilst many bear the costs.

Equality depends on property rights. Biodiversity is a resource and property rights are the ability to secure use and derive value from the resource. There are four main types of ownership rights: **open** and unregulated, **common**, regulated by rights and responsibilities, **private** and **state**. These systems mesh with legal systems which can be either **custom and tradition** or **formal legislation**. Another factor influencing equity is **social position**. For example, many clashes in the developing world pit indigenous smallholders against outsiders and large companies. No one property system has a monopoly of virtue. The important factor is security of tenure, bringing a sense of responsibility to future generations. This permits a long-term view rather than short-term cut and run. The rights of future generations, inherent in many definitions of sustainability, have proven especially difficult to build into economic processes (see Plate 18).

Plate 18 Nature reserve wardens intercept illegal logging in Mozambique. Classic human pressure on local resources. The situation does not look tense because the wardens are well aware that local people need wood fuel. Management policies which recognises the needs of local people and the potential for customary, sustainable use may offer a better option than "fortress and fence" reserves.

Cultural attitudes

The diversity of attitudes to biodiversity is astonishing but three broad cultural patterns exist. Many traditional societies combine use with a sense of nature's importance to sustain the society, so that rights, responsibilities and taboos effectively create rules which conserve the natural system. These cultures are termed **ecosystem people**. Conversely **biosphere people**, typically in the developed world and urban elites of developing countries, treat biodiversity as a resource which can be exploited to destruction. Resources are drawn from a global catchment and new resources can be opened up once others are exhausted. Biosphere cultures may lose the sense of rights of future generations in the expectation that new resources will always be found. The ecosystem to biosphere culture shift spawns a third category, **ecological refugees**. They are ecosystem people forced to move to new sites. Shanty towns of many developing world cities are the obvious example but historic ecological refugees include the early European colonists of North America and settlers on the western frontier of the United States (see Box 21). Ecological refugees exploit the resource base but often have little regard for their environment, lacking local knowledge or sense of security into the future. This is a recipe for degrading biodiversity.

Institutional failures

Institutional failures include institutional weakness and lack of knowledge.

Institutional weakness

All the good intentions and treaties in the world are ineffective if the will and means to conserve biodiversity are lacking. Many developing countries, with fragile political structures vulnerable to coercion by force or corruption, are vulnerable. However, even

---

Box 21

Human impacts: economic and social forces – the passenger pigeon and American buffalo

The extinction of the passenger pigeon (*Ectopistes migratorius*) and near extermination of the American buffalo (*Bus bison athabascal*) are apocryphal examples of species loss. Both cases show the interplay of economic and social factors.

Culture

Both species occurred in proverbially vast numbers on the frontier of the expanding United States, an area contested by the increasing settler population (ecological refugees) and native Indian cultures (ecosystem people).

Inequality and property rights

The pigeon and buffalo were an open access resource for settlers, whilst inequality of power undermined the property rights of the native Indians. Buffalo hunting was used to undermine the native Indian cultures, which were reliant on the buffalo as a resource and spiritual focus.

Economic forces

Commercial hunting of pigeon and buffalo went on until the 1880s, when the last large scale buffalo hunt took place in 1884. Economic forces sustained the hunts even as numbers declined. The expansion of the railroad network improved access to distant stocks and movement of carcasses and products, so that demand could increase whilst prices remained low. Buffalo hunting was encouraged by failures of other leather supplies and improved tanning technology. The market was also easy to enter and leave, allowing hunters to switch in and out depending on the availability of other jobs. Ultimately conventional economics suggests that extinction should still be avoided as populations become so low the costs of hunting become too great relative to value. In neither case did this happen and prices remained broadly stable, even in the last year of hunting. Evidence of local employment also suggests that buffalo hunters did not anticipate any final collapse until it had happened. There were professional pigeon hunters even after wild stocks were gone. The buffalo was saved by the establishment of eighteen herds, most fewer than ten animals, between 1873 and 1919. The passenger pigeon lingered on in at least three captive zoo stocks, including at least twenty in Cincinnati Zoo, where the last one died.

---

well-established schemes can be undermined, for example the fate of Project Tiger in India. Set up in 1972 Project Tiger co-ordinated a system of reserves focused around the tiger (*Panthera tigris*), and also benefiting many other species. Initially successful estimates of Bengal tiger numbers rose from 1800 to 4300 in 1992. Political upheaval, particularly the deaths of Prime Ministers Indira Ghandi and Rajiv Gandhi (in 1984 and 1991 respectively), started a rapid collapse. In 1993 numbers fell to 3750 as patronage decayed and habitat was threatened by new policies to open up the Indian economy to foreign companies. Funds dried up and poaching began, even in famous tiger reserves. Dealers in tiger products repeatedly evaded imprisonment. By 1995 only 2500 tigers were left, with estimates of one tiger killed every day. A survey of Project Tiger schemes revealed that 80 per cent lacked any armed anti-poacher patrols and 75 per cent did not receive funds on time. In October 1995 a meeting in Delhi set out to revive the work: 'Save the Tiger, save the jungle, save India'. Corruption, greed, intimidation and indifference are mighty powerful enemies.

Lack of knowledge

Our incomplete inventory of life inevitably endangers biodiversity. What expertise exists is unevenly spread, concentrated in the developed world. Lack of knowledge can even threaten well-known species of the developed world.

The extinction of the Large Blue butterfly (*Maculinea arion*) in Britain is a classic case. First recognised in the eighteenth century in colonies scattered throughout southern England, the Large Blue showed local extinctions by the 1900s and, despite intense conservation work, the last colony was lost in 1979. Half the known sites had been lost to direct habitat destruction but many other protected colonies had also been lost. Too late we had understood the butterflies' requirements; its life history was worked out only in 1915. The caterpillars mature as parasites inside ants' nests. Several species of ant will pick up young caterpillars and carry them back to a nest. In the 1970s it became clear that one ant species, *Myrmica sabuleti*, is a much better host from the caterpillar's point of view. Other ant hosts attack the caterpillars more readily. *Myrmica sabuleti* thrive in swards with some grazing. Without grazing the tall vegetation cools the microclimate, so other ant species dominate. Intense grazing reduces the grass sward so yet other species move in. By this time the last few Large Blue sites had had grazing halted as a conservation measure or, where rabbits were present, myxomatosis reduced their impact. Our ignorance had doomed the Large Blue. In 1983 European stock was reintroduced but the unique British race was no more.

Economic failures

Economic failures include market failure and intervention and subsidies.

Market failure to capture full value of biodiversity

This problem afflicts both the most affluent northern hemisphere and the poorest developing economies. The mechanisms are similar even if the detail of habitats and species varies around the world. It is also perhaps a less obvious threat than some of the glaring problems of local politics or agricultural change, but though less tangible it is all the more dangerous and to some extent drives the other problems.

Markets do not capture the full value of biodiversity to society, whether global or local. Generally an individual person or an organisation will gain financially from exploiting resources now, given the uncertainty of the future. Natural capital, including biodiversity, is turned into manmade financial or material capital. Not only does the individual gain the immediate profit but also very often many of the costs (e.g. effluent) can be dispersed into the environment so the full cost of exploiting the resource (e.g. effluent treatment or containment) is not carried by the individual beneficiary. A problem shared is a profit increased. In the mean time society at large ends up paying for the burden of any such costs, perhaps by direct, society-wide payments (e.g. taxes to support anti-pollution management and treatment) but more insidiously through loss of environmental quality and ecosystem function. Such damage can accumulate and its effects become apparent only when damage has been done. The so-called free market is even freer than supposed since individuals can dump some of the costs of their activities onto everyone else. Such gratis get-outs are called **externalities**, because the costs are borne by others. The benefits of intact, healthy biodiversity are missing from economics.

The individual does lose out too, suffering the same loss of environmental quality as everyone else, though financial security may buy some protection from environmental degradation. The short-term gain may be incentive enough to compensate. Individual versus global costs and benefits from exploitation diverge so biodiversity is lost. The problem may be exacerbated where the general benefits give little local return or even inflict costs. The former is exemplified by demands for conservation of the tropical rainforests, largely found in developing countries, to maintain atmospheric quality in the industrialised developed world.

Intervention and subsidies

Institutional policy can drive the loss of biodiversity. Interventions are often seen as politically easier than promoting economics that fully value biodiversity but even well-intentioned interference can be damaging.

Direct subsidy of destructive activity occurs worldwide. Not only does biodiversity suffer from undervaluation but also such policies heap on additional bias in favour of exploitation. The environment is degraded and sustainable economics undermined. The global total for such destructive subsidies has been estimated at $600 billion per year. Examples include subsidy of intensive agriculture in Europe encouraging habitat destruction and subsidy of forest clearance in Brazil for livestock. Subsidies continued to rise during the 1980s. Proportions of farmers' incomes made up from subsidy in different continents in the periods 1981–84 compared to 1989–92 included Australia 11 versus 12 per cent, European Union 32 versus 46 per cent and the United States 27 versus 27 per cent. Subsidies are rife in the developing world, often in support of damaging practices, for example the percentage of pesticide price subsidised in 1993 was Senegal 89 per cent, Ecuador 41 per cent and Indonesia 82 per cent. Resources used for subsidy are also diverted from more sustainable uses such as education and research, thus dooming deprived populations to exploit more land using unsustainable techniques. Since the early 1990s increased awareness of damage done by direct subsidy of destructive activities has resulted in a change in policy, e.g. revision of tax subsidy for clearance of Brazilian forests for cattle ranching.

Wildlife in wartime

Faced with the human tragedy of war, a concern for the wildlife and habitats caught in the fighting seems almost inappropriate. However, the impact of human conflict on biodiversity came to prominence during the 1990s, partly because of conflicts that have affected conservation flagships, such as the mountain gorillas on the Rwanda–Congo border made famous by Dian Fossey. In 1999 an attack by militia on a gorilla-watching camp resulted in eight tourists and a warden being killed, attracting global media coverage. Governments and humanitarian agencies have realised the role of biodiversity as a key element for sustainability and human well-being and the need to conserve these assets as far as is possible even whilst conflicts rage. Current wars are often better defined as **complex emergencies**, often within rather than between states, combining low-level fighting between local militias, civic and economic collapse and threats associated with natural disasters, such as starvation or disease, and migration, for example in the conflict in the Congo or West Africa. During these wars people often rely on local biodiversity for vital resources and in some places modern protected areas coincide with ancient refuges, for example some of the montane forest of the Udzungwa Mountains in Tanzania. Humanitarian agencies and governments now recognise that functioning ecosystems are crucial for post-war recovery. The Red Cross and UN High Commissioner for Refugees (UNHCR) now include biodiversity as one key to recovery.

Impacts of conflicts on nature vary. Protected areas, often inaccessible, become front lines used by one side or the other and destroyed in fighting or in a scorched earth policy. Low-level conflict in parts of Colombia has been credited with keeping down human populations, resulting in less exploitation of forests, although sites targeted for drug cultivation may be illegally converted from pristine forest. This can include targeting of animals seen as economically valuable to the other side, for example a possible threat to the mountain gorillas. The chaos of war may allow illegal exploitation to develop, e.g. logging or exploitation of bushmeat (see Box 20). At the very least the insecurity will undermine any sense of long-term ownership and benefit, encouraging short-term exploitation. Protected areas may be damaged as people use their resources or move in to escape attacks. A lack of management and general debris and pollution degrade the habitats. It is not just humans who flee. Zambian farmers have blamed increased damage from elephants on the political strife in neighbouring Zimbabwe, resulting in elephants moving across the border to get away from now vulnerable Zimbabwean protected areas.

Biodiversity can sometimes benefit. Fighting, or the aftermath such as minefields, may create no-go areas where any exploitation, legal or otherwise, is suspended. However, this no man's land benefit has diminished in recent complex emergencies which lack any clear frontlines. Combatants may protect valuable natural resources if they have the means to. Bitter experience has created ground rules for attempts to conserve biodiversity during conflicts. Resilience can be strengthened prior to conflicts by promoting local ownership of resources, the adoption of peace parks, supporting staff working in protected areas and raising awareness amongst humanitarian and peace-keeping agencies. During conflicts any work anticipates the value to biodiversity rebuilding society. The value of illegal exploitation of biodiversity by combatants can be reduced by trade embargos, parks staff may be supported and maintained and management of refugees might include strategies to mitigate destructive degradation of biodiversity. Even combatants may be open to being induced to avoid destroying sites and species, thus avoiding bad publicity and gaining prestige. After the conflict, priority should be given to restoring whole functional ecosystems, rebuilding parks staff and protection so that wildlife and resources cannot be rubbed out in any post-conflict vacuum and integrating conservation and economic rehabilitation. These actions not only help in the short term as civil society recovers but also provide longer term economic benefits which may alleviate some of the stresses leading to conflict.

Away from the complex emergencies of the developing world, military structures have provided some fine wildlife habitat. Porton Down, established in 1916 as the UK research establishment for chemical weapons, is the single most important site for butterflies in the United Kingdom, boasting forty-five species, 80 per cent of the national list. In the United States, the Denver Rocky Mountain Arsenal, a dump opened up for chemical weapons during the Second World War, and full of toxic waste, is now a prime wildlife site.

Valuing biodiversity: economics for conservation

We benefit from biodiversity in many ways. In 2001 the UN launched the Millennium ecosystem assessment to explore how changes to ecosystem services may affect our well-being, and how best to respond to these threats, feeding into the Johannesburg Plan of implementation. Importantly this project recognised the variety of benefits, not only the obvious raw materials or services but also how these affect freedom, health and security. The UN started with core ecosystem services, such as nutrients, soils and primary production. These in turn provided three classes of direct service: first, **provisioning services**, such as food and water, second, **regulating services**, such as climate and disease, and third, **cultural services**, such as spiritual and educational ones. These three services all provide part of four constituents of well-being: **security** (e.g. clean and safe shelter), **basic material** for a good life (e.g. resources for livelihood), **health** (e.g. nourishment) and **good social relations** (e.g. cultural expression). All of these four constituents need to be fulfilled to create freedom and choice. Biodiversity as fundamental to freedom and choice? We have come a long way from pure biology.

Methods for costing value of biodiversity

There are many ways in which we value biodiversity. This section deals primarily with the methods used to put a financial value on gene, species and habitats, in effect a **utilitarian** approach. There are at least three other approaches: **social-cultural** systems, which give value through ideas of identity, taboo and belief, **ecological value**, expressed as the security, resilience and health of the ecosystem, and **intrinsic value**, the innate worth of something regardless of what it does or who says. Using methods that apply a monetary value to biodiversity does not appeal to everyone. There can be tensions between different value systems. The Judeo-Christian tradition described in Genesis is widely interpreted as giving humanity mastery over the natural world; animals, once made, were declared 'good' by God, suggesting an intrinsic worth, but how much is that? Many cultures regard particular species as the spirits of ancestors and are quite reasonably averse to putting a price on their great-grandmother. Native cultures may harbour contradictions. For example, for Aborigines in Australia, places have intrinsic values but most individual animals and plants do not.

The true economic value of biodiversity is a new field in which some possibilities are emerging. If the true value of biodiversity can be properly realised and included in economic mechanisms, then a very powerful tool for conservation is available. First, we should distinguish the types of value that biodiversity may represent.

1 **Use value**

- direct use, e.g. harvesting, tourism
- indirect use, e.g. ecosystem function
- future use, e.g. potential uses, insurance against the unknown.

2 **Non-use value** (the benefits of not using biodiversity). Essentially the amount we are **willing to pay** to conserve, **willing to accept** in compensation for not exploiting or **willing to forgo** by not maximising a financial return, to keep the biodiversity. Such concepts include:

- **bequest value**, the classic 'our children's children' argument
- **existence value**, willingness to pay to conserve a species which we may not even see, content to know it survives
- **option value**, potential uses, direct, indirect and insurance against the unknown.

Several approaches have been used to put financial values to these concepts.

Changes in productivity and economic gain

These are direct attempts to value economic returns associated with sustainable management compared to degradative exploitation.

Contingent valuation

Contingent valuation is the establishment of a financial value based on willingness to pay, willingness to accept or willingness to forgo. Values based on the willingness to accept a reduced return often exceed those based on willingness to pay to conserve. Willingness to accept is an abstract loss of future returns which are not in the bank, whilst willingness to pay involves handing over what has already been earned. The term contingent reflects the severe biases that can ruin such valuations, commonly based on surveys and questionnaires.

Hedonic pricing

Hedonic pricing relies on the expertise of economists to tease out the value of natural phenomena and environmental quality from within the total price of a resource. For example a house in beautiful, unspoilt British countryside would fetch a different price from exactly the same house in the middle of a grim urban conurbation. If this separation can be achieved, general models of demand and value can be built. This is tricky and requires skill and sufficient data on prices. The technique works only for products that are being marketed so that an overall price exists which can be broken into components.

Travel costs

The value of biodiversity as a recreational resource can be estimated from the costs that tourists are willing to pay to visit sites and view wildlife. Some costs may be easy to specify (e.g. air fare or park entrance fee), but holidays are often not solely to see wildlife. Some costs can be difficult to specify. In addition the damage done by tourists, whether through resources used or uncosted externalities, should be accounted for. In some cases the income can be expressed directly as a financial value for the land area involved, allowing comparison with potential alternative land uses.

Valuation of substitutes

A technique used to value ecosystem functions relies on costing the technology that would be needed to replace degraded biodiversity. One role of wetlands is as a sink for excess nutrient run-off, protecting rivers from such degrading effluent. The cost of drainage infrastructure and machinery necessary to achieve a similar diminution in nutrient pollution can be calculated directly.

The economics of conservation

Since the mid-1980s, there has been an explosion of work trying to tie in the intuitive feeling that biodiversity is valuable with financial measures of this worth. Although techniques are tentative, produce differing estimates and depend on often patchy data, the economic arguments for conservation have shifted from hazy notions to startling estimates of the financial value of biodiversity.

Genetic resources

The Green Revolution of the 1960s and 1970s raised productivity of wheat and rice by some 60 per cent; this was partly due to genetic improvements to cultivars. As production rose so did fluctuations in yields, in part due to the narrow genetic base of crops used. In the 1970s only four varieties made up 75 per cent of the US potato crop, an echo of the Irish potato famine of 1846, when reliance on one variety that was devastated by potato blight led to 1 million deaths and emigration of a further 1.5 million. Genetic uniformity has resulted in crop vulnerability throughout recent decades e.g. in the United States: 1950s wheat rust, 1970s maize corn leaf blight, 1984 Florida citrus bacteria infections. Genetic diversity in crops is an insurance against the vulnerability of monocultures and may also reduce need for chemical supplements, cutting costs and environmental damage. The genetic diversity is also the raw material for manipulations. Crop varieties and livestock breeds can be improved for particular conditions, for resistance to pests and disease, to boost productivity and product quality and to aid cultivation techniques. The US tomato industry received a multimillion dollar benefit by inclusion of a jointless stalk gene found in the wild only on the Galapagos Islands. Estimates of the value to agriculture can be made in several ways. An example of comparison of costs for research versus resulting gain come from corn breeding in the United States. In 1984, $100 million was spent on research but $190 million earned as a result. Improved crop yields are an obvious measure. In the 1980s US total crop productivity rose due to new cultivars by $1 billion. In Asia rice and wheat productivity in the Green Revolution by $1.5 billion to $2.0 billion.

Genetic diversity is highly valued for pharmaceutical products. Plant products can be used directly, or as raw material to refine into therapeutic derivatives, or as the inspiration for synthesis of artificial analogues. In the last case the value of biodiversity is as pure information. Estimates of the value of plant-based pharmaceuticals include some extensive work in the United States where plant products were used in 255 drugs prescribed between 1953 and 1973. The plant components were estimated as worth $1.6 billion, rising to $9.8 billion in 1980, then $18 billion in 1985. Since the bulk of products came from only 40 plant taxa the 1953–73 estimate is equivalent to broadly $200 million per species. The value can be doubled when hospital use is included. Add to this the savings from illness avoided or cured and therefore work days not lost and estimates of the annual benefit in the mid-1980s were between $34 billion and £300 billion. These data have been taken further. Estimates suggest that 5000 species of plant had to be examined to find these 40 most useful (i.e. 1 in 125). Assuming the same proportion holds globally (we are out on the thin ice of assumptions here) for every 1000 species of plant extinct, 8 could have yielded very useful drugs. The return to pharmaceutical prospectors from new drugs found in the Costa Rican forests have been cited as $4.8 million per drug. In this light the loss of economic options from plant extinctions are high and the retention of option values from conservation valuable. Advances in **biotechnology** and **bioassay** have spurred the practical prospecting for new plant-based drugs such as Taxol, an anti-cancer drug derived from the bark of the Pacific yew (*Taxus breviflora*). In recent years rights to prospect for genetic biodiversity have attracted substantial deals. In 2000 the pharmaceutical company Merck paid the Instituto National de Biodiversidad in Costa Rica $1 million for the rights to 2000 samples, with follow-up royalties for any resulting commercial product, whilst in 1997 Diversa, a Californian company, paid the US National Parks Service $175,000 for rights to conduct research on heat-resistant bacteria from Yellowstone Park's thermal springs.

Taxonomic diversity

Many species have a financial value from direct use, e.g. as a sport target or tourist attraction. Their value is often directly estimated from financial returns linked to sport or tourism (see Box 22). Species also have a vital but still poorly understood role in the function of ecosystems, which is much trickier to estimate. Ecosystems may depend on a small set of **keystone** or **driver** species, the other taxa present **passengers**, an excess sometimes

Box 22

The value of conservation in the United Kingdom

The Royal Society for the Protection of Birds (RSPB) has valued the benefits from conservation to local UK economies, revealing a major economic driver, which is particularly important in some of the more far-flung, economically disadvantaged regions. The RSPB estimate that there are the equivalent of 18,000 full-time jobs in conservation throughout the United Kingdom. In addition to this direct employment conservation projects draw in money from tourists. The RSPB's Boat of Garten Osprey Centre in Scotland pulls in £1.7 million a year to the local economy, supporting 87 jobs in the area. Overall, visitors to RSPB reserves spent £12 million a year, supporting 200 jobs on reserves and 30 jobs locally. Local effects may be important in peripheral regions, such as 100 jobs in mid-Wales, a rural region, based around the red kite; £1.3 million a year is spent in Orkney by wildlife tourists and £6 million a year from birdwatchers in North Norfolk supports an estimated 135 jobs. New conservation schemes are promoted in part for their economic impact. In 2004, 30 red kites were released on the fringes of Gateshead in Tyne and Wear. As the publicity put it, 'they will be a valuable new visitor attraction for the area, helping generate income for the local economy'.

The value of wildlife may provide opportunities to resolve conflicts. In the United Kingdom there is growing presure from sea fishermen for a cull of seals. Seal populations have been increasing and fishermen suggest that seal predation lowers fish stocks. A study in Cornwall, where seal damage to netted fish has been costed at £100,000 per year, gave estimates of public 'willingness to pay' to see grey seals between £8 and £9, with a conservative valuation of seal viewing at £526,000 per annum. If some link can be made between losses to fishermen and income generated, then seals more than pay their way. A similar situation exists around the Farne Islands, which hold up to 3 per cent of the UK grey seal population. Wildlife tourism based on boat trips out to the Farnes was estimated as worth £1.5 million a year in the 1990s, a huge boost to a comparatively remote part of England (see Plate 19).

described as **ecological redundancy**. However, passenger taxa may become important if the environment changes. The link between species and critical ecosystems function is contentious but the functional diversity of species, what they do rather than taxonomic relatedness, may be important for ecosystem resilience. This resilience is the size and frequency of disturbance that an ecosystem can withstand without substantial change. Such roles are very difficult to cost.

A financial value can be put on direct use species. Estimates of the value of the African elephant in the late 1980s range between $22 million and £30 million per annum, based on several different measures of costs borne by tourists. Similarly the value of Kenya's protected areas, their attraction largely due to species, was estimated as $540 million in 1994, greater than economic returns from any alternative use. Attempts have been made to value individual species, often relying on the willingness-to-pay method to assess how much individuals would be willing to contribute towards the conservation of species. Here are some examples: in the United States monk seal and humpback whale $9.6–16.8 and whooping crane $21–141, whilst in Sweden, wolf $100–130. I do not know of any schemes where the general public has then been asked to hand over the money.

Plate 19 Grey Seals (*Halichoerus grypus*) on the Farne Islands, north east England. The UK supports internationally important breeding stocks of the Grey Seal, but their conservation has become politically sensitive due to the belief of some fishermen that the seals take too many of the dwindling North Sea fish stocks.

Ecosystems

Like species, ecosystems can be valued both for direct gains and indirect benefits, particularly ecosystem functions. Central to all valuations is the comparison of benefit from conserving an ecosystem versus the costs of management and of opportunities lost. The financial value of tourism has proven an effective approach. Global ecotourism in 1988 was estimated as worth $233 billion, of which half could be linked directly to biodiversity. Detailed studies of tourism to Costa Rican rainforests in the early 1990s suggested that their value was $1250 per hectare, against costs of conservation management of $30–100 ha and value of adjacent farm land of only $30–100 ha. Conservation would be economically the best option, so long as the benefit reached local people. Valuing ecosystem functions has proven trickier. Functions comprise **regulation** (e.g. climate, watershed processes), **production** (e.g. food, fuel, oxygen), **carrier** (e.g. space for agriculture) and **information** (e.g. spiritual, scientific roles). The ecosystems functions create resources or carry out processes with substantial economic benefits.

Most attention has focused on tropical forests and attempts made to measure the **total economic value** which must capture the direct, indirect, option and existence values. Even when this is not possible, values can be put to different aspects of their benefit.

Value versus alternative uses

Even if complex, often intangible values are difficult to pin down, comparisons to alternative uses can be revealing. A classic study of the Korup National Park, Cameroon, estimated benefits from use, function, trade and tourism of £23.6 million versus set-up costs and lost opportunity of £16 million, a net gain of £7.6 million. Similarly a study of Bacuit Bay in the Philippines compared benefits from multi-use including logging to those without logging, tourism and fisheries on an adjacent coral reef. Note in this case the explicit benefit to another ecosystem, the reef. Depending on precise estimates of future returns, the benefits of not logging over ten years were between $11.5 million and $17.5 million.

Local use values

Many studies from around the world have assessed the value of local sustainable use of forest. The range of values is considerable but many are high. South American examples (all given as value in dollars per hectare per year) include an experimental Caiman harvest in Venezuela, 0.75, Peruvian villagers' multi-use of forest resources, 16–22, and wildlife value of Ecuador rainforest, 120. Indian examples include forests as a source for domestic elephants, 3.0, village household multi-use, 50, and India-wide gross benefits from local use of food and medicine plants, 117–144.

Individual ecosystem functions

Estimates for the value of specific functions have concentrated on the role of forests as a carbon store. At a global level one hectare of tropical forest has been valued as $200 and if cleared, making contribution to global warming via carbon release of $2000–4000. Some **externality**!

The Amazon as a whole, acting as a carbon store, has been valued at $46 billion, which, when combined with a direct use value of $15 billion and existence value of $30 billion adds to a total value of $90 billion. At a local scale estimates of the value of rainforests, from different sources, using different techniques, include Costa Rica $102–214, Thailand $400, value to United States of foreign rainforests, $500 (all as value per hectare per year).

So long as economics can capture the total value of biodiversity, in particular externalities and ecosystem functions and ensure that local people benefit, economics can be a very powerful incentive tool for conservation.

The total value of the biodiversity services

In 1997 the attention-grabbing publication of an estimated value for the world's ecosystem services was US$33 trillion. The estimate arose from an intensive workshop, pulling together experts in ecological economics who compiled data from over a hundred individual studies to give the value of services provided by major biomes. The idea was simple: estimate the value per hectare for each biome, multiply this up by the area of each biome and add the results to get a global total. For this study ecosystem services were defined as goods and services. Non-renewables such as fossil fuels were excluded and the value of the natural capital (i.e. the infrastructure of the forests, oceans and other ecosystems) was not included because this is effectively infinite since they cannot be replaced. The study used seventeen major services such as climate regulation and waste treatment; these were deduced for sixteen biomes, e.g. open ocean and tropical forests. The value of each biome for each service was expressed as US$ per hectare per year. Table 4.3 gives summary data for selected biomes and services.

When all the data were compiled the average value was US$32,268,000,000,000 per year, the figure often quoted as $33 trillion. This is 1.8 times greater than the global annual gross national product, a widely used measure of the sum total of economic activity. We could not afford to replace biodiversity even if we had the technology. Varying some of the assumptions gave a range of between $16 trillion and $54 trillion. The $33 trillion estimate is likely to be a minimum. The estimate missed out some services and some biomes for which no reliable data were available.

The figure has attracted criticism. The multitude of studies from which data were drawn used a variety of valuation methods, such as contingent valuation or willingness to pay,

Table 4.3 Summary data for the value of selected ecosystem services. The four services (climate regulation, waste treatment, nutrient cycling, recreation) and the annual value of one hectare of the biome are all given as US$ ha^{-1} yr^{-1}. The total global value of the biome for all services and total value of an ecosystem services across all biomes are given as US$ year^{-1} × 109. Question marks (?) = insufficient data to estimate value. Note the variation in biome value. Freshwater wetlands are not a major biome in extent but are highly important for their services

Biome	Climate regulation	Waste treatment	Nutrient cycling	Recreation	Annual value of one hectare biome	Total global value of biome for all services, US$ year^{-1} 10^9
Open ocean	?	?	118	?	252	8,381
Tropical forest	223	87	922	112	2,007	4,706
Grasslands	0	87	?	2	232	906
Freshwater wetlands	?	1,659	?	491	19,580	3,231
Total value of ecosystems' services across all biomes, US$ year^{-1} × 10^9	684	2,277	17,075	815		

Source: Constanza *et al.* 1997

so it was assumed that the data are reliable and comparable. Even if the data are robust the estimate itself may be a very blunt instrument, with little sense of priorities and threats. The estimate also ignores the ability of many systems converted to human use, e.g. agriculture, to provide services. Another useful measure would be how much better (or worse) do natural ecosystems perform as providers than converted systems. Comparisons of the total economic value, which combines any local gains to individuals from benefits such as crops with global benefits such as ecosystem services, of largely pristine ecosystems versus land converted for human use suggests that conversion results in lower value. For example the value of Canadian wetland, expressed as year 2000 US$ per hectare until 2050, is $8000, compared to $3700 if converted to farm land, whilst that of Thailand mangrove is $60,000, versus $20,000 if converted. So why do we keep turning natural ecosystems into farmland? The global benefits are largely ignored by current market systems, there is often a short-term gain to the individual upon conversion and we lack sufficient information about the consequences of long-term conversion for so many habitats. Global budgets for conservation of protected areas in 2002 were estimated at around $6.5 billion and between $20 and $28 billion would be needed to compensate people for not converting remaining protected areas for the next thirty years. Over the same time period these areas would provide an estimated $4000 billion to $5200 billion benefits, a benefit–cost ratio of 100:1. A good investment.

Some environmentalists have also attacked this approach and indeed the whole market valuation of biodiversity in general. Valuation has been proposed to help make decisions to maximise benefits to society, resolve conflicts and allow trade-offs. However, monetary valuation may not capture the full value of biodiversity. How can you measure naturalness in economic terms? What is the value of a diversity of antelopes? The use of monetary value has also been promoted as the language of policy-makers and markets, intended to catch their attention. Should we risk reducing biodiversity to just another commodity? One criticism of attempts to use market forces to promote sustainability (e.g. the ability of countries and companies to trade carbon pollution quotas) is that the rich and powerful, often in the developed world, will be able to buy these rights from the poorer, developing world, and therefore carry on polluting, whilst the developing world exceeds its remaining quotas which cannot be enforced for want of infrastructure or political will. Many environmentalists see the market place as a monster to be met with a very long spoon.

Local economic incentives and conservation

Some of the most effective progress linking biodiversity and economics has developed to create benefits for local people from conservation of wildlife. This is best illustrated by two schemes, one a project founded by a voluntary conservation body focused on a special site in Cameroon, the second is Zimbabwe's national programme linking wildlife and local people, CAMPFIRE.

The Mount Kupe Forest Project, Cameroon

Mount Kupe is a granite peak, swathed in rainforest, covering a mere 21 sq. km. Designated as the Mount Kupe National Park, it is a biodiversity hot spot famed for its birds. There are at least fourteen endemic species, most famously the Mount Kupe bush shrike (*Malaconotus kupensis*), thought to be extinct until its rediscovery in 1989. In addition the grey-necked picathartes is described as a cross between a crow and vulture that skulks around in caves. There are at least 300 other bird species in the park. The area was conserved in part by local taboo as the source of wealth and well-being, but outsiders now outnumber indigenous locals and the park is under pressure from clearance for banana plantations and hunting meat.

In 1991 BirdLife International started the Mount Kupe Forest Project following preliminary meetings with the local villagers. The project intends to conserve the montane habitats by facilitating economic benefit to local people from the presence of intact forest, based on income from tourists who stay with families in the adjacent village of Nyasoso (see Plate 20).

The first tourists came in 1991–92, with 41 visitors staying a total of 160 tourist nights. In 1992–93 numbers rose to 94, staying 440 tourist nights. The income to the village was estimated as £5500 in total. In 1994 the cost of a night's stay per person was £6, with additional income to locals acting as guides. This money goes straight to the villagers. In addition tourists' donations to the project raised £323, of which 40 per cent was used on trail and new campsite management, with 60 per cent going to the village. Bird-watchers have proven the most valuable visitors, staying more nights to see more birds (see Plate 21).

There are many additional benefits on top of the cash. An education programme includes awareness of conservation. Pride in the project is promoted by noticeboards displaying foreign publicity. A pen pals scheme, run by the Royal Society for the Protection of Birds, linked local children to UK schools. A rickety charabanc library circulates material to more schools and has become a much loved icon for the children. Visitors are encouraged to talk to children about the importance of the birds and to let them try their binoculars and telescopes. The scheme trains teachers and produces a teaching magazine; adult education includes agricultural and health advice. Some locals have become very involved with the

Plate 20 Mount Kupe Forest Project, Cameroon. Publicising the benefits of conservation. A local building sports a picture of the Mount Kupe Bush Shrike, a flagship species for conservation of the site.

Plate 21 Mount Kupe Forest Project. Visiting birdwatchers are asked to talk to children about their enthusiasm for wildlife and its importance and to let them have a go with binoculars and telescopes.

project, gaining experience from trips abroad and training in ecological monitoring so that skills are transferred into the village. The project even sponsors a football team.

The CAMPFIRE programme, Zimbabwe

One of the most famous national schemes to link conservation of biodiversity with economic benefit was the Communal Areas Management Programme For Indigenous Resources (**CAMPFIRE**). Before colonisation the black population had utilised wildlife through systems of common ownership carrying rights and responsibilities. These links were broken in the colonial era and much of the local population forced to live in tribal trust lands (latterly named communal areas), which were on economically marginal terrain. The CAMPFIRE scheme reinstates control by and benefits to local people, creating an incentive for conservation of biodiversity.

CAMPFIRE's aims are long-term development, management and sustainable utilisation of natural resources in communal areas (CAs), management of resources by placing responsibility and custody with local people, allowing communities to benefit directly from exploitation of resources in their CA and providing administrative back-up and advice. Alternatively, and better capturing the spirit of the scheme, Zimbabweans talk of the choice between wildlife as relic or resource, of wildlife paying its way: 'those who pay the social cost of living with wildlife should reap the economic benefit'. The costs can be high with crops destroyed, livestock lost and people killed every year by dangerous game such as elephants. CAMPFIRE schemes vary in detail of wildlife involved and utilisation (meat, live capture for sale, big game hunting, photographic safari, fishing). They are developed through a protocol linking local people to the Department of National Parks and Wildlife Management (DNPWLM) working with district councils.

A typical early example is the Dande Communal Area CAMPFIRE, begun in 1987. Dande CA is in north-east Zimbabwe, wedged between a safari area and other communal areas. Soils are poor, rainfall erratic and tsetse fly, which transmit sleeping sickness, occur. Ironically tsetse control in the 1980s saw a 30 per cent rise in local populations without any real improvement of the economic base. In 1987 the DNPWLM advised locals on options and opportunities, helped with initial administration and surveys of wildlife to draw up quotas. The Dande Council co-ordinated local wards. The northern half of the communal area was leased to a commercial big game hunting safari, the southern half used to set up a hunt owned and managed locally, to boost local skills. Problems arose with initial investment from campsites and lack of marketing, so grant aid was provided by the Zimbabwe Trust. In 1989 the first revenue rolled in. The local owned safari raised Z$299,000, but cost Z$232,000, so profit was Z$6700. The private company proceeds added Z$168,000 and grant aid Z$98,000. The total profit of Z$333,000 was split amongst 7000 people, about Z$47 each. This income is substantial. The statutory minimum wage in Zimbabwe was Z$240, but in rural areas local people earn much less. Income per household from other CAMPFIRE schemes at that time ranged between Z$3 and Z$2306, the latter enough to buy 38 goats (see Figure 4.8).

There remain problems, often associated with distributing such largesse, but the major problems have been linked to international attitudes to conservation. The Dande CAMPFIRE scheme is not untypical in its reliance on big game hunting. In 1993 big game hunting in Zimbabwe was worth US$4 million, of which elephants made up 70 per cent. Killing elephants is politically fraught: the Zimbabweans and other southern African countries have been pushing to reopen the ivory trade. Opposition has been fierce. Legal ivory trade might open up poaching and laundering of ivory, wiping out remnant elephant populations elsewhere in Africa. Elephants have also gained status as charismatic, sentient creatures beloved of TV audiences to whom the very idea of shooting them is abhorrent.

Many people think of wildlife
as a pest or problem . . .

but . . . .

Hunting
Safari hunters pay large sums
of money to shoot wild animals.

Figure 4.8 CAMPFIRE programme literature aimed at Zimbabwean villages to promote conservation of wildlife by utilisation as a sustainable resource

Source: Zimbabwe Trust.

Box 23

'Poaching is wiping out Zimbabwe's wildlife': Zimbabwe Conservation Task Force 2004

By the late 1990s Zimbabwe's conservation work, combining superb protected areas with CAMPFIRE to open up local participation and benefits, was heralded as an example to the whole world. By 2004 Zimbabwe had become a superb example of how institutional collapse can undermine conservation and endanger species, driven by land occupations, economic collapse and a breakdown in law and order. It is difficult to assess the impact on wildlife, partly because doing so risks being seen as seditious by the Zimbabwean government, but here are some examples and general assessments.

The Wildlife Estate, including the main protected areas, has been undermined by loss of staff, lack of funding, poaching and some occupations. Black rhino have been poached in main reserves such as Hwange: the population across all Zimbabwe fell from a late 1990s estimate of at most 500 to about 200. A major initiative to create a huge international trans-boundary reserve combining Gaza, Kruger and Gonarezhu (in Mozambique, South Africa and Zimbabwe respectively) is threatened because the Zimbabwean preparations have collapsed and part of the reserve is occupied by squatters. By 2004 only 12 out of 88 private game conservancies remained unoccupied. In most conservancies wildlife has been decimated by poaching and occupation of these smaller sites. The situation has bred corruption. Wildlife rangers have been accused of taking bribes to allow illegal big game hunts. In 2003 South African hunters shot a black rhino. In the Mufurudzi Safari Area senior wardens have been accused of working alongside government supporters and poachers to drive out tourists, allowing poaching and illegal game hunts. Whilst general tourism has collapsed, big game hunting appears less affected: there are still plenty of websites advertising tours. There have been increasingly desperate and ad hoc responses. WWF arranged for 22 black rhino to be translocated away from areas of poaching. The Born Free Foundation has set up its own anti-poaching unit concentrating on areas near Hwange where a few African hunting dogs remain.

ZimConservation is a new group of conservationists and citizens attempting to provide some assessment of the crisis via a website. Meanwhile the CAMPFIRE website, http://www.campfire-zimbabwe.org, is now reduced to a page on trout fishing.

Advocates of the Zimbabwean approach have been accused of a mixture of naivety and self-serving careerism. Shooting wildlife to conserve it remains a battleground of competing attitudes (see Box 23).

Conservation's targets

Many Zimbabwean mammals have an explicit price on their head: the official minimum trophy fee a hunter must pay to shoot one. Initial 1991 prices, in Zimbabwean dollars, probably under valued the resource, e.g. bull elephant Z\$10000 or a warthog at Z\$200, equivalent to £1000 or £20. Trophy hunting has survived the political crisis in Zimbabwe, having always operated in a very private world. With the Zimbabwean dollar all but worthless trophy fees are now quoted in hard currency, e.g. a bull elephant fee in 2005 is US\$10,000, or for those who have not seen Disney's *The Lion King*, a warthog at US\$350.

However political and economic collapse may allow the system to be abused, for example at one site US$50,000 bought the hunters as many targets as they wanted of anything. When the CAMPFIRE system was working effectively, the animals were money on the hoof to local people, targets for hunters, with the hunters themselves targets for specialist holiday companies. This is not the image that many people have of conservation and there are campaigns to halt the entire trade, even when regulated (e.g. Bloody Business). The campaigns often focus on so-called 'canned hunting', where animals such as lions are kept in enclosures and shot much as people might shoot at gonks in a fairground stall. There are more familiar and generally less contentious approaches to conservation, treaties, reserves, **reintroductions** and biodiversity. However, given that the threats to biodiversity are so firmly rooted in human pressures and economic failures, the examples of Mount Kupe and Zimbabwe are lessons we should not ignore. Conservation has undergone a sea-change in recent years. It is no longer simply a matter of protecting animals on nature reserves but of conserving and utilising biodiversity in all its forms, as Chapter 5 will show.

Summary

- Current extinction rates are difficult to estimate but many techniques converge on estimates suggesting that losses are much higher than natural background rates.
- Ecological patterns suggest that there are general processes which increase vulnerability to extinction and cause final extirpation.
- Accelerated losses are due to human pressures of population growth, culture, institutional and economic failure.

Discussion questions

1   How many species of vertebrate have been driven to extinction in your country and with what consequences?
2   Devise an economic value for a common wild animal or plant in your region.
3   Is extinction an ethical problem?

Further reading

See also

Natural extinctions, Chapter 1 pp32–37.
The ecology of biodiversity, Chapter 2 pp47–69.
The legislative framework, Chapter 5 pp174–184.

General further reading

Barbault, and Sastrapradja, S.D. (1995) *Generation, Maintenance and Loss of Biodiversity*. In V.H. Heywood (ed.) Section 4, *Global Biodiversity Assessment*, Cambridge University Press, Cambridge.
Detailed review of the ecology of extinctions.

Gaston, K.J. (1994) *Rarity*. Chapman and Hall, London.
Fascinating review of rarity, definitions, patterns and processes.

Lawton, J.H. and May, R.M. (1995) *Extinction Rates*. Oxford University Press, Oxford.
Detailed review of extinction rates and risks, past and present.

McNeely, J.A., Gadgil, M., Levegue, C., Padoch, C. and Redford, K. (1995) *Human Influences on Biodiversity*. In V.H. Heywood (ed.) Section 11, *Global Diversity Assessment*. Cambridge University Press, Cambridge.
Detailed review of causes and consequences of human pressures on biodiversity.

Pearce, D. and Moran, D. (1994) *The Economic Value of Biodiversity*. Earthscan, London.
Short but sweet introduction to possibilities and problems for economic valuation of biodiversity.

Thomas, C.D., Cameron, A., Green, R.E., Bakkenes, M., Beaumont, L.J., Collingham, Y.C., Erasmus, B.F.N., de Siqueira, M.F., Grainger, A., Annah, L., Hughes, L., Huntley, B., van Jaarsveld, A.S., Midgley, G.F., Miles, L., Ortegaa-Huerta, M.A., Peterson, A.T., Phillips, O.L. and Williams, S.E. (2004). Extinction risk from climate change. *Nature*, 427: 145–148.
Some recent climate change and habitat loss estimates of global extinction risk.

Other resources

Asian vulture crisis

http://www.biordlife.net/action/science/species/asia_vulture_crisis/

Zimbabwe

Ad-hoc crisis management by Born Free Foundation.
http://www.bornfree.org.uk/elefriends/zimbabwe.htm
A new Zimbabwe monitoring site, Zimbabwe Conservation.
http://users.starpower.net/aeveans123/
The sad debris of the CAMPFIRE site
http://www.campfire-zimbabwe.org/

Up-to-date reports on WWF site

http://www.pands.org/news_facts/newsroom/

Anti big-game hunting campaign 'Bloody Business'

http://www.bloodybusiness.com/

5   The conservation of biodiversity

Conservation means much more than guarding charismatic species inside fenced reserves. This chapter covers:

- **New attitudes to conservation**
- **Legislation, treaties and funding**
- **Protected areas**
- **Biodiversity in captivity**

Conservation as a recognisable, coherent scientific movement coalesced from diverse, piecemeal actions and insights following the Second World War. Attempts to secure the scientific understanding and management of species and habitats for conservation saw the foundation of expert organisations at both international (e.g. WWF) and national levels, such as the United Kingdom's Nature Conservancy, set up in 1949. In the developed world this concentration on scientific management of species and sites has been very successful, at least in conserving rare species. In the United Kingdom many once rare or extinct species of bird such as the osprey (*Pandion haliaetus*) or avocet (*Recurvirostra avosetta*) have made remarkable comebacks. Being rare and charismatic and in the United Kingdom is a good recipe for survival. Meanwhile common species such as the house sparrow have undergone severe declines, whilst outside of the developed world the concentration on protected areas and scientific management may be altogether inappropriate.

Since the 1980s conservation has undergone a sea-change, driven in part by the multi-faceted nature of biodiversity. Conservation has become much more than the province of scientists. Conservation science and environmentalism swapped insights. In particular conservationists appreciated that the factors driving habitat loss and species extinctions lay outside the immediate horizons of biology. The ultimate human impacts were the product of economics, politics and society. Conservation expanded its horizons with the realisation that protection of biodiversity, whether individual genes, species or ecosystems, could not be achieved by science alone, however expert or however committed. Conservation could not work in isolation. Integration of conservation into the wider realms of economics and politics was vital. Max Nicholson, the influential originator of so much UK conservation, analysed the progress of ecology and conservation and pointed out that the task is reconciling three apparently intractable elements – people, area and the biosphere. The science of biodiversity and the mysteries of ecology and evolution were easy compared to explaining the lessons to societies that did not want to know. The conservation of biodiversity became part of broader social, economic and political concerns. The science of conservation had to combine with campaigns for public awareness, participation and sustainable economics.

Evolving concepts for biodiversity conservation

Conservation science's conceptual framework shifted, even at the risk of science being submerged in wider topics. The United Nations Environment Programme summarises the need for this shift (see Figure 5.1).

- to integrate different approaches to ensure the widest possible range of biodiversity is conserved
- to recognise that conservation is heavily influenced by social, cultural, economic and political factors
- to encourage co-operation and co-ordination of policy and institutions.

The conservation of habitats and species in the sense of direct, science-based management is described as the **protectionist** approach. This strategy had its roots in European and American attitudes to nature which combined the privileging of scientific expertise to manage and control nature with a Romantic idyll of wilderness in which people were an intrusion. This is a comparatively recent phenomenon. Prior to the Enlightenment, wilderness was not some mythical Eden but a dangerous place, hence the name from old northern European languages more or less translated as 'wild beast's lair'. Following the romanticisation of wild places, natural ecosystems and the local people who lived in them were seen as unruly and dangerous. This attitude is evident in the colonial era as European administrations created protectionist nature reserves in tropical countries, often removing indigenous people from the land. The same attitudes applied closer to home. The history

The people must be able to:

1. Manage community involvement and participation.

2. Identify water sources.

3. Identify land use areas (areas for wildlife, grazing, crops, people).

4. Protect natural resources.

Figure 5.1 Zimbabwean CAMPFIRE publicity comic reflecting conservation's widening scope, combining protection of biodiversity, sustainable use and local ownership of resources

Source: Zimbabwe Trust

of land management in Scotland follows a similar pattern with local, communal ownership swept away during the Highland clearances, people driven off the land and replaced by large estates owned by gentry and used for sheep ranching. These were in turn supplanted by estates run for recreation, bought up by wealthy industrialists and aristocrats, aping the style of Queen Victoria and the romanticised Highlands of her Balmoral retreat. Local people were excluded just as much as people driven from protected areas in the colonies, their livelihoods linked to the landscape at best through menial, insecure employment. The first African protected areas were designated at the same time as the first UK reserves, e.g. Sabie in South Africa in 1892, Wicken Fen in England in 1899.

The schism between land, livelihoods and people was reinforced by the privileging of a technocratic science as the model for management, which ignored local knowledge, separated nature from culture by the emphasis on designations (e.g. Sites of Special Scientific Interest, National Nature Reserves), focused on species and ecosystems whilst ignoring the role of people in the landscape and created a bureaucracy that could be portrayed as distant and colonial. An example is the undermining of the UK Nature Conservancy Council, with its headquarters in south-east England, during the height of battles over reckless forestry plantations in Scotland in the 1980s. The conservation profession that grew in the post-war developed world faced a serious critique, that indigenous people were either seen as a problem or romanticised, that the expert bureaucracy distrusted the general public (and vice versa) and was consumed by procedures, that for the developing countries to bear the burden of conservation whilst the developed world carried on with business as normal looked like hypocrisy and that conservationists were still driven by a vision of unspoilt nature as in some sort of equilibrium from which all human impacts had to be removed. Protected areas became fortress conservation, 'fines and fences' an echo of imperial legacies, combining a romanticised nature with echoes of the days of the exclusive hunting party.

Conservation now means much more besides. The World Conservation Strategy has three explicit components: **protection, sustainable use** of biodiversity and **sharing the benefits** of this use. Article 1 of the Rio Biodiversity Treaty echoes this with its remit as 'the conservation of biological diversity, the sustainable use of its components and the fair and equitable sharing of the benefits'. Protection of biodiversity is now allied with promoting sustainable exploitation. The developing world has led the way, beginning with participatory or **community-based conservation**, building approaches which recognised that local people must be involved with and benefit from conservation schemes if protected areas were ever to be more than a burden. This approach caught on extensively throughout Africa where weak governments and lack of money undermined protected areas, whilst exploitable wildlife and a heavy dependence on international aid encouraged some economic exploitation of wildlife. The change recognised that protected areas were not effective in many parts of the world, that local people should be involved and that sustainable development might be able to use wildlife.

Many such schemes appear to have been successful, though this new trend has its own problems. Linking conservation with sustainability, woolly notions of community involvement and the emphasis on using wildlife all played well to social-political agendas dominant from the 1990s. There are constraints. Zimbabwe's flagship CAMPFIRE scheme has provided not only successes but also salutary lessons. Not all areas may contain sufficient, exploitable biodiversity so disillusion sets in. Revenue generated from wildlife may simply replace funds that central government withdraws and successful schemes may attract the unwanted attention and involvement of powerful elites. Many CAMPFIRE schemes have worked in the short term but the exploitation of wildlife may just be seen as a stop-gap, with no real acquisition of skills by locals. Areas with successful schemes may attract people, increasing pressures on the land. In the worst cases some schemes failed because they were seen as a top-down imposition on people, who were obliged to participate – not a recipe for sustainable conservation.

Reviews of sustainable use of wildlife in Africa also suggest that economic benefits are unclear. Lump-sum payouts may be useful for big projects, such as building a school or hospital, but may not meet day-to-day needs, and the money might not equal income that could have been generated by using the land for something other than wildlife, so-called opportunity costs. In Kenya the estimated opportunity costs of wildlife have been estimated at US$203 million per year, versus a return of US$27 million. These schemes will work only if benefits exceed all costs, provide day-to-day livelihood and ensure that no members of the community are worse off.

Recently negotiated strategies have been developed, which recognise that participatory schemes still limit local involvement to participating in someone else's agenda, very often the only participation being the occasional visit from a government official to tell people that wildlife was a good thing and hand out some money. Good negotiated schemes involve setting priorities for sites which may involve more than just conservation goals, for example schemes might recognise rights of access and use of some resources. The negotiations should define a wide social definition of aims, combining conservation of wildlife with other needs. Conservation should, it is hoped, become integral to all activities, rather than the risk with participatory systems which, at their worst, reduce nature solely to an economic resource to be exploited. Good examples of negotiated schemes have been developed in Tanzania in the 1990s, in and around the Eastern Arc Mountains, a biodiversity hot spot which includes the Udzungwa mountain range where the partridge of Chapter 2 was discovered. Here protected area and participatory approaches had failed, with locals still not feeling any sense of ownership or benefit from forest reserves, and still seeing themselves as 'thieves of the forest' even during participatory schemes. Since 1998 local negotiation for access and use has been developed. This not only includes obvious use of products (e.g. fruits, honey and timber), but also recognises the function of sites as resilient sources of clean water, for example, and the significance of sites as special places in local cultures. Negotiated priorities should be every bit as important in the developed world too. The problems of eking out a living in the Highlands of Scotland are as fraught as in the plains of Africa, with the revealing twists that landownership in Africa often makes conservation empowering local people easier to promote.

Context

National policies and attitudes to conservation will affect the methods used. Success will depend on recognition of this and integration of conservation with national social, economic and political goals. The United Kingdom and Zimbabwe are good examples of such differences.

The United Kingdom has a strong tradition of statutory and voluntary protection. The statutory mechanisms were organised by the National Parks and Access to the Countryside Act 1949, revised as the Wildlife and Countryside Act 1981. The concentration on protection by use of protected areas is evident from the diversity of designations (see Figure 5.2).

- *National Nature Reserves (NNRs):* national, often internationally important sites, owned or managed by agreement by statutory conservation bodies. There are over 330 in the United Kingdom.
- *Sites of Special Scientific Interest (SSSIs, or Areas of SSI in Northern Ireland):* originally set up as counterparts to NNRs to provide regulation of valued sites outside the NNR system. They are now seen as the main national mechanisms to safeguard

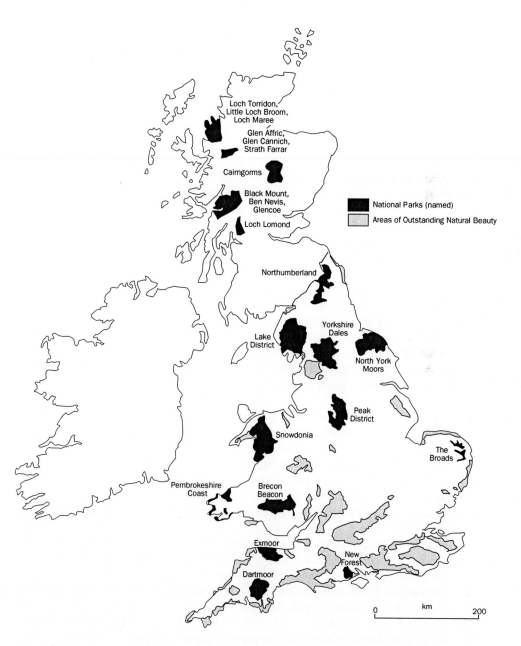

Figure 5.2 Extent of two UK protected area systems. (1) National Parks (and their equivalent in Scotland). Multi-use protected landscapes encompassing the total area within their boundaries. Governing bodies have some restrictive powers and an increasing primary purpose of nature conservation. (2) Areas of Outstanding Natural Beauty, landscaped recognised for their natural and aesthetic quality. No single managing authority but within the area landowners may apply for aid to manage in an environmentally sensitive manner

sites, many outside formal reserves. There are over 6500 in the United Kingdom, with over 8.0 per cent of land area.

- *Special Areas of Conservation (SACs):* sites selected, in response to EC Habitats Directive, to maintain conservation value at European level. In the United Kingdom 605 are proposed, many as areas embracing several individual sites assessed using SSSI criteria. The sites embrace 76 habitat types and 40 species out of the European wide list of 169 habitats and 623 species for protection.
- *Special Protection Areas (SPAs)*: designated for conservation of breeding and migratory birds in response to EC Birds Directive. There are over 240 in the United Kingdom. (SACs and SPAs are intended to form a pan-European network of sites: Natura 2000.)
- *Marine Nature Reserves (MNRs):* much like National Nature Reserves but extending out to the limit of territorial waters. There are only two so far.
- *Local Nature Reserves (LNRs):* valued sites designated by local authorities; over 900 in the United Kingdom.

In addition to this protectionist reserve network, biodiversity in the United Kingdom benefits from a host of other area designations that provide some regulation or incentive for good management: National Parks, Areas of Outstanding Natural Beauty and Environmentally Sensitive Areas.

Zimbabwe's history has created a different structure (see Figure 5.3). Responsibility for all wildlife is vested in the Department of National Parks and Wildlife Management, the infrastructure based on the Parks and Wildlife Act 1975. The Act also set up a Parks and Wildlife Board to advise ministers on conservation and utilisation of wildlife. Conservation is focused on the Wildlife Estate, over 13 per cent of Zimbabwe's area (see Plate 22). The Estate is solely state-owned land or land held in trust by a statutory body. The Parks and Wildlife Act 1975 also recognises the work that can be done on private land and requires the Department to enhance wildlife management and production in all categories of land. Note the duality: conservation is linked with utilisation and production.

The Wildlife Estate consists of five categories:

- *National Parks:* intended to conserve the natural landscape, wildlife and ecology. Whilst access for tourism and recreation is permitted, the variety of activities is limited and sport hunting is excluded. There are currently twelve in number.
- *Safari Areas:* conservation of habitats and wildlife but with the express purpose of providing opportunities for public access for tourism and recreation, including sport hunting. Safari areas may be almost as valuable habitat as National Parks and are often adjacent. The difference lies in the policy of more extensive exploitation. There are seventeen.
- *Sanctuaries:* areas set up to protect animals for the enjoyment and pleasure of the public. Sanctuaries are often smaller than National Parks or Safari Areas, supporting less diversity of species. There are twenty-six.
- *Recreational Parks:* nineteen areas set up for conservation of natural features for the enjoyment and recreation of the public.
- *Botanical Reserves and Botanical Gardens:* set up for the conservation of rare and indigenous plants and botanical communities. Seventeen exist.

There are additional national designations, e.g. Cloud Forest Region, National Monument and State Forest reflecting other management priorities

The shifting culture of conservation has had an impact. In 1994 the UK government published *Biodiversity: The UK Action Plan*, drawing together existing policies, practice

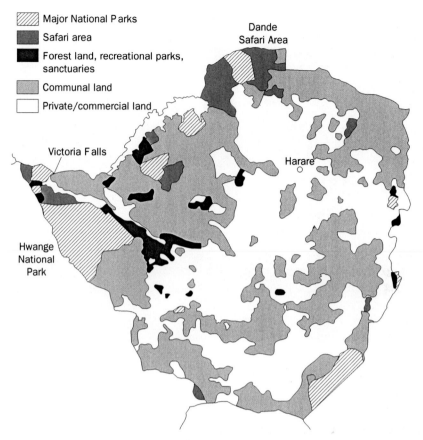

Figure 5.3 Extent of Zimbabwean protected areas. National Parks, Safari areas and sanctuaries are all part of the Parks and Wildlife Estate controlled by the Department of National Parks and Wildlife Management. Communal Areas may take up the CAMPFIRE programme. The Dande Safari Area described in Chapter 4 is indicated as well as Hwange National Park (plate 22) and Victoria Falls, Crocodile Farm (plate 23)

and strategy for the first time. This was a direct response to Rio's Article 1, an attempt to fuse ecological principles with economic and sustainability. Preliminary work driven by the Joint Nature Conservancy Council (the government's national umbrella) and the Department of Environment (a government ministry) identified four themes: conservation of resources (i.e. protection, establishment and management of areas), sustainable use, community involvement (from individuals through to government agencies) and data collection and information. The UK Action Plan refined this into six aspects: conservation **in-situ**, **ex-situ**, sustainable use, partnership and education, UK support for biodiversity overseas, information and data. A steering group then took the UK Action Plan forward, producing *Biodiversity: the UK Steering Group Report. Volume 1* in 1995 with detailed plans including:

- costed targets for protection of species and habitats
- accessibility and co-ordination of biological databases
- increase public awareness and involvement
- mechanism for implementation and review of the Plan.

The need for a dynamic, updated, adaptive response is emphasised throughout. *Volume II* contains lists of key species, key habitats and broad habitat goals, costs and responsibilities.

Plate 22 Elephants (*Loxodonta africanus*) in Zimbabwe at a water hole in front of a viewing platform in Hwange National Park. The impressive adults and playful babies are a fine sight but for local people they represent a direct threat to crops and life.

These include 116 key species and 14 habitat action plans plus reviews of the status of 37 main habitat types found in the United Kingdom. Species action plans identify responsibility for actions, rationale for targets, legislative protection, site protection, monitoring, review protocols, international co-operation, threats from air quality, climate and development of land. Habitat plans and statements include definitions, principal threats, rationale for targets and the broader context within which costed Action Plans can be drawn up. Boxes 24 and 25 summarise a Species Action Plan for Desmoulin's Whorl Snail and a Habitat Action Plan for Fens. The impact of biodiversity on conservation, with the horizons broadened beyond solely protection, had for the first time produced a national, integrated species and habitat protection plan.

The legislative framework

International treaties

Conservation laws developed as piecemeal, national legislation. International legislation conceded that sovereign rights were paramount and seldom recognised the importance of links in natural systems spilling over political boundaries. The Stockholm conference not only reiterated the importance of sovereign rights but also promoted responsibilities of states not to be a source of damage beyond their borders. The Rio Convention was the first international treaty to include legal obligations to act on this responsibility.

International treaties for conservation can work not only by imposing direct conditions but also through creating a supportive culture and incentives. Adoption of a treaty signals

Box 24

A Species Action Plan for Desmoulin's whorl snail

Desmoulin's whorl snail (*Vertigo moulinsiana*), routinely described as 'about the size of a breadcrumb', became a conservation icon during the 1995–96 anti-roads protests started against the building of the Newbury bypass in southern England. The protests involved direct mass action to obstruct construction work. As the protests reached their height, it was suddenly realised that many of the wetlands across which the bypass would be built were home to this diminutive snail, prompting pariliamentary questions. At the same time the snail appeared as one of the 166 Species Action Plans. The Action Plan summarised: **Current status**, **Factors causing loss**, **Current action**, **Action Plan objectives** and **Proposed action** (*Policy and legislation, Site safeguard and management, Species management and protection, Advisory, Future research and monitoring and Communications and publicity*).

Given the No Action proposal for publicity, Desmoulin's whorl snail attracted intense interest, television news coverage, newspaper articles, even a question on the BBC's prestigious *Any Questions?* programme. An unfortunate representative of English Nature interviewed by the BBC resorted to stating that our lack of knowledge about the snail meant that its presence could not justify holding up the construction of the bypass, a reversal of the precautionary principle that probably sounded as foolish to him as it did to the audience.

The bypass is now open and surveys throughout southern England suggest that the snail may not be as rare as once thought. A UKBAP report suggests that targets have been met to carry out surveys which have shown a wider distribution than previously recorded, raise public awareness and positive media coverage and produce a summary leaflet on status and ecology.

See this snail and find out more at http://www.arkive.org/species/ARK/inverte brates_ terrestrial_and_freshwater/Vertigo_moulinsiana

The snail is a species listed as worthy of SAC site designation: see http ://www. jncc.gov.uk/ProtectedSites/SACselection/species.asp?FeatureIntCode =S1016

Box 25

A costed Habitat Action Plan for the Fens

The Action Plan summarised: **Current status**, **Threats**, **Current Action** (*Legal status, Management*), **Action Plan targets** (*Identify priority sites in need of, and initiate, rehabilitation by 2005. All rich fen and other sites with rare species should be considered. Ensure appropriate water quality/quantity for all SSSI/ASSI fens by 2005*), **Action required** (*Policy and legislation, Site safeguard and management, Advisory, International, Future research and monitoring, Communication and publicity*) and **Costings** (*the successful implementation of the action plan will have* *implications for the public and private sector*). The Action Plan lists estimated costs to maintain 1200 hectares through to 2010 (per annum); 1997, £40,000; 2000, £70,000; 2010, £70,000). Two clear objectives are given: initiate restoration of priority sites by 2005 and ensure appropriate water quality and quantity for all SSSI/ASSI fens by 2005.

A progress report shows that a full inventory has been produced for England, although Scotland, Wales and Northern Ireland lag behind. In most areas fenland shows a decline, albeit slowing. Both targets (restoration of critical sites and water quality improvement) are behind schedule, with a lack of specific targets in terms of water quality and quantity needs hindering raising the profile of this BAP. Some site specific successes are outlined.

Find out more about UKBAPs for habitats at http://www.ukbap.org.uk/ habitats. aspx.

---

national support for conservation, perhaps a nudge to national legislatures to adopt specific laws. International treaties recognise global benefits and local costs of living with biodiversity, especially important to those developing countries harbouring the greatest diversity but least able to afford conservation or avoid destructive exploitation.

Building treaties

The first wildlife conservation treaty, the Convention for the Protection of Birds Useful to Agriculture, was signed in 1902. Since then over 150 environmental treaties have been signed, but lack of co-ordination, gaps and duplication impede their operation. Most treaties develop as a negotiated text adopted by countries but come into force only once a country consents, usually by signature and ratification (or accession for later entrants). The treaty may come into force only when all participants, or an agreed threshold, consent. The success of a treaty typically requires review, update and administration to ensure compliance, through a secretariat and regular conference. Five of the global treaties prior to Rio have been the most influential for conservation. These are:

- Ramsar Convention on Wetlands of International Importance. First signed in Ramsar, Iran, in 1971, it is now adopted by 146 countries.
- World Heritage Convention. The Convention Concerning the Protection of the World Cultural and Natural Heritage was signed in 1972, and is currently adopted by 180 countries.
- Convention on International Trade in Endangered Species of Wild Flora and Fauna (CITES). The first version was signed in 1973 and is currently adopted by 166 countries.
- Convention on the Conservation of Migratory Species of Wild Animals, first signed 1979 and, by 2005, adopted by 90 countries.
- The Convention on the Law of the Sea (UNCLOS) has been ratified by 148 countries.

International treaties: mechanisms

International treaties use up to four approaches to conservation: **funding**, usually for national efforts on behalf of a global resource; designation of **protected areas**; regulation of **biotechnology rights** especially in defence of local people; **regulated trade**, whether outright prohibition, restricted exploitation and mechanisms to ensure financial rewards to locals.

Funding

Funding schemes recognise the obligations of richer countries towards the developing world, so often required to conserve the bulk of biodiversity. Examples include the World Heritage Fund, linked to the World Heritage Convention (WHC). The WHC has proven a popular treaty due to the availability of finance and has an annual budget now in excess of US$4 million. The Global Environment Facility grew from an IMF–World Bank Development Committee meeting in 1989, to finance protection of the global commons. Increasingly the GEF is a device to fund new international treaties. The GEF took on the financial administration of Rio Convention moneys, until a 1995 Conference of Parties (i.e. those countries that had signed the Convention) could agree on a mechanism. Since 1991 the GEF has granted US$4.5 billion, provided by 32 donor countries, for 1300 projects in 140 countries, including protection of the ozone layer, control of greenhouse gas emissions and conservation of biodiversity; this funding is being matched by an additional US$14.5 billion from other partners. A review of GEF funded projects for biodiversity between 1992 and 2001 reported that US$1.8 billion of GEF funding plus additional US$2 billion from partners had been targeted on the conservation and sustainable use of biodiversity, across 395 projects in 123 countries, the majority linked to new or existing protected areas, of which the majority were forest ecosystems. Half the projects had achieved most of their objectives, many therefore had not. In particular there were problems judging the impact for want of baseline data and many projects lacked clear planning for long-term sustainability, including effective monitoring, once the funding was finished. The report spotlighted the need for longer term, smaller money projects, linked to economic growth and security, involving local communities (who were only heavily involved in 30 per cent of projects), build local capacity and expertise to manage the ecosystems and addressing the root causes of biodiversity loss. It would be easy to criticise, but mean spirited, the GEF represents a major contribution to the conservation of biodiversity, increasingly rooted in the wider social and economic context.

Protected areas

The benefits of international treaties requiring signatories to designate protected reserves work via reciprocal rewards. A country may incur costs or forgo opportunities to establish reserves within its borders but benefits from similar sites set up in other states.

Biotechnology rights

Organisms, their parts and products, were typically excluded from treaties regulating patents, ownership and exploitation rights but biotechnology and the potential of genetic diversity have propelled biotechnology rights to the forefront of international conservation law. In particular there is concern as to how to ensure **source countries**, so often the developing world, benefit from exploitation of their biodiversity which is refined in the **recipient countries**, mostly the developed world.

Until the Rio Convention, international treaties did allow some exclusive rights over plant varieties and other organisms. Plant varieties could be registered with the Protection of New Varieties of Plants (1961). Patents, designed for rights over new and useful inventions, were extended to organisms in 1980 in the United States as the result of a court case, Diamond versus Chakrabarty, over a genetically engineered bacterium. The Rio Convention included no specific mechanism for patenting biodiversity although four Articles talk of exchange, fair and equitable sharing of benefits and co-operation, especially to help developing countries, a steer against restrictive patenting.

Tensions remain. Accusations of biopiracy have been fuelled by revelations such as those in 1996 that an American biotechnology company, Phytera, had signed contracts with several European botanical gardens by which the gardens provide tissue specimens from their collections. In return the gardens will receive a share of any profits from products developed. Unfortunately some of the contracts omitted commitments to share some of the profits with the countries of origin of the plants.

Lessons have been learned and there are fine examples of how to build equitable partnerships, allowing biodiversity and local, traditional knowledge to be used whilst sharing the benefits, in line with the exhortations of the Rio Convention. There is a need for both prior consent and agreements of how any products might be used and the profits distributed. The identities of those involved need to be clearly stated, which resources are to be used and who owns them: ownership, exclusivity and confidentiality all need to be clear from the start. Schemes are better developed for pharmaceuticals and include royalties paid on sales, initial fees and milestone payments as the product is developed and non-monetary benefits such as collaborative research, training and medical assistance. Some companies, such as Glaxo, and universities, e.g. Berkeley, have published their own protocols.

International trade in wildlife

The total value of wildlife trade (whole organisms, parts and products) comes as a shock. Although hard to be precise, estimates suggest annual values of timber US$100,000 million to $130,000 million, fisheries US$40,00 billion to $50,000 billion, mammal furs US$750 million and ornamental plants US$250 million. The edible snail trade alone is worth US$460 million a year. There is a dark side to this, with the illegal component trade worth between US$5000 billion and $8000 billion a year, the second largest illegal business sector after drugs. Legal trade is regulated by the CITES, originally a prohibitive treaty, forbidding all trade in taxa listed in Appendix I, trade by permit only for Appendix II species and regulation of trade in species requested by countries that have national regulations to prevent unsustainable use or illegal exploitation and need other countries to help (Appendix III). The CITES also requires signatures to provide annual records of Appendix II trade, to monitor trends and check against estimated stocks. CITES' emphasis has shifted. The 1981 New Delhi Conference of Parties permitted change of some Appendix I species to Appendix II for sustainable use, primarily ranching. The status of the African elephant is a good example of local variation, reflecting the status of the species and potential for sustainable use in southern Africa. This is an Appendix I species except for populations in Botswana, Namibia, South Africa and Zimbabwe, which are listed as Appendix II, with country-specific caveats. Zimbabwean elephants may be used legally for non-commercial export of hunting tropics, live export to appropriate destinations, export of hides and export of leather goods and ivory carvings of non-commercial purposes. In contrast the other three countries may trade raw ivory within strict limits of size, weight, site of origin and verification. The first example of variation was the Nile crocodile in Zimbabwe in 1983. A management quota system, relying on countries to devise quotas and consumer countries to enforce compliance accepting only those imports with valid documents, was permitted. These opportunities brought problems. National quotas could be poorly set, either by ignorance or design. Even if a country's quota was reached, wildlife could still be harvested, smuggled over the border and laundered into stocks of adjacent states. Global trade in crocodile products has been a test of the two competing arguments for legitimate trade, either that legal trade will extinguish illegal trade which is no longer needed or lucrative, or that legal trade will act as a Trojan Horse encouraging illegality with the laundering of illegitimate products into the markets (see Plate 23). Of the twenty-three species of crocodiles and their relatives, fifteen are commercially valuable. Until the 1970s most

Plate 23 The author wrestling with a would-be Zimbabwean hand-bag at a crocodile farm at Victoria Falls. The farm hatches eggs laid in captivity and 5 per cent of two-year-old crocodiles are released into the wild. The rest are turned into clothing, ornaments or meals.

products were the result of unregulated hunting, including poaching from protected areas. In the 1980s CITES regulations came into force, including generic global regulation and targeting specific countries, e.g. closure of legal trade to close loopholes. Trade in crocodile products has grown substantially. Comparing 1977 with 1997 numbers of crocodilians from captive breed, ranched or wild stocks were for 1977, 0, 1258 and 38,831, and for 1997, 46,249, 257,248 and 74,955 respectively. Unregulated exploitation of wild populations has been replaced by regulated farmed stocks. Of the twelve species used for commercial exploitation, eleven are the least threatened of all crocodilians.

There are currently some 827 Appendix I species, 4466 animals and 29,074 plants in Appendix II, and 291 species in Appendix III; however, whole families can be listed, so the inclusion of the orchid family adds up to 35,000 species.

The Convention on Biological Diversity

Rapidly evolving biotechnology and trade treaties reflect the ecological, economic and social origins of biodiversity. The Rio Convention displays this mix of conservation by obligation, equity rights and positive incentives. Access to genetic resources by foreign collectors requires prior consent by source countries, fees and royalties and mechanisms to share the outcomes of research. The transfer of technologies for conservation to the developing world not only is encouraged but also requires preferential terms. This equity and access by source countries and people override patent laws. Essentially conservation in the developing world is explicitly tied to obligations on the developed world, for expertise and money, despite the best efforts of some signatories to write opt-outs into the Convention treaty (Box 26).

Box 26

Rio Convention caveats

Articles 20 and 21 of the Rio treaty deal with financial resources and financial mechanisms respectively. Paragraph 1 of Article 20 states, 'each Contracting Party undertakes to provide, in accordance with its capabilities, financial support and incentives in respect of . . . the objectives of this Convention', followed by paragraph 1 of Article 21, 'There shall be a mechanism for the provision of financial resources to developing country parties'. The United Kingdom's official publication of the treaty includes this revealing caveat at the end: 'The government of the United Kingdom . . . declare their understanding that the decisions to be taken by the Conference of Parties under para. 1 of Article 21 concern "the amount of resources needed" by the financial mechanism, and that nothing in Article 21 authorises the Conference of parties to take decisions concerning the amount, nature, frequency or size of the contributions of the Parties'. France and Italy added similar nervous notes.

National Legislation

National legislation differs in detail between countries but common themes recur. Laws can be categorised into two approaches: first, **regulatory**, involving prohibition or restriction, backed by punishment versus non-regulatory relying on positive incentives, and second, the focus for laws on genetic resources, species, habitats or destructive activities.

Regulation

Genetic resources
Legislation to conserve genetic resources is a recent development, with little precedent to build upon and an indicator of changed attitudes to conservation. Protection is typically combined with regulation of exploitation and sharing the benefits. Regulation of genetic biodiversity is founded in the Rio Convention's statement of the sovereign rights of nations to control access and use of genetic resources, vividly revealing the tensions between source and recipient countries. Source countries face three tasks: regulation of collection, including sharing information and participation of local people; regulations for sharing technologies and benefits; establishment of regulatory authorities to oversee compliance. In addition source countries will succeed only if recipient countries develop complementary legislation to allow enforcement.

Species protection
Traditional custom, taboo and law protected some game animals in many countries, with restrictions by season, catch, method or entitlement. The majority of species went un-protected, often held to be the property of the landowner. The first species-specific law was drawn up in the Swiss Canton of Zug in 1911, to regulate collection of the edelweiss (*Leontodon alpinum*). Article 8(k) of the Rio Convention calls for countries to develop legislation for the protection of threatened species.

Species-based legislation commonly lists chosen species, banning or limiting interference. The taking and possession of wildlife are often linked to bans on trade. Since the act of collecting often goes unseen, the prohibition of possession, whether of whole specimens or parts of products, is the best control, whilst trade restrictions limit financial rewards.

Legislation can also require conservation and recovery schemes, again for a species or habitat, financial incentives and, rarely, the precautionary principle that any collection or alteration of habitat must be proven to cause no harm. The Sri Lankan Fauna and Flora Protection Ordnance of 1993 is an example of the latter.

Problems for listing species arise from lack of criteria, evidence or expertise. Many national lists are short, dominated by the inevitable spectacular vertebrates. Legislation is easier to draw up for state-owned land. Elsewhere restrictions on the rights of private landowners may be regarded as an infringement of liberty and cause resentment.

Protected areas

The bland jargon 'protected areas' covers a host of legislation from nature reserves, characterised by prohibition or restriction of activities within designated sites, habitats or regions and works in many ways. UNEP recognises over a thousand different designations worldwide.

Specific **nature reserves** owned by government or voluntary conservation bodies are the most obvious. State control of land can be problematic, requiring compulsory purchase or first refusal to buy and, if these are acceptable, may be limited by money. Management can be impeded by the need for buffer zones, external impacts that cannot be escaped (e.g. atmospheric pollution) and the human problem of integrating local interests with what might be seen as the heavy hand of government or outsiders.

Conservation can be achieved without the rigour of a protectionist reserve. Many countries designate **wilderness areas** that forbid and regulate certain activities by landowners and visitors, the land use controls perhaps zoned allowing some activities in parts but not throughout. Other approaches include **site-specific regulation** of privately owned land (e.g. UK SSSIs) or are **habitat specific**, granting general protection to a habitat type throughout a country, whether an ecosystem such as wetland, or a more particular feature such as a pond. **Planning controls** covering all land can be useful, especially inclusion of less valuable landscape, e.g. urban. These controls are either comprehensive, anticipating developments, perhaps through **zoning**, or reactive, dealing with applications on a site-by-site basis.

Regulation of damaging activities

Many human activities degrade biodiversity, their impact spreading far beyond the source. However well established a protected area is, these outside activities can be a serious threat and national legislation often seeks to limit such damage. There are two main problems, activities within a protected area that unintentionally cause damage and those causing damage beyond their source site. The former includes the impacts of walking, climbing and vehicle scrambling. In such cases the very quality of the natural habitat is the lure. The second category includes pollutants, pesticides and the release of exotic species. Damage from either type is difficult to regulate. Legislation often results from damage already done. Damage on private land is especially tricky to control; incentives for good management may be more effective.

Non-regulation: incentives and agreements

Regulation can be ineffective, stymied by lack of means or will to enforce, the clash of legislation versus individual rights and likely to cause resentment. **Positive incentives** to conserve are increasingly used. There are three main forms. Financial schemes include incentives to conserve such as tax reductions or exemption for conservation management schemes and abolition of incentives to destroy. Restrictive agreements require a landowner to manage a site in a specified way. Such agreements include **covenants**, which work in perpetuity so that subsequent owners must maintain the work, and **easements**, an obligation often linked to a very precise task or action. These agreements sound coercive but can allow conservationists to sell land which can be used in some ways so long as the specific goals are met. **Management agreements** use contracts to start or maintain positive management in return for payment. The payments can be linked to a particular practice, such as low intensity farming or creation and maintenance of specified habitats.

International aid

The Rio Convention explicitly stated the need for financial aid for conservation from the developed to the developing world. Richer countries were asked to back their presidents' speeches with money. Financial aid is a vital complement to laws and treaties.

Development aid

Biodiversity can benefit from **unilateral development aid**, i.e. from one country directly to another, but specific allocations to conservation are difficult to pinpoint in general aid budgets. Of the sixteen countries in the OECD Development Assistance Committee in 1988, only six could identify funds allocated specifically for biodiversity, though others were moving to this. In addition to direct funds screening other aid projects for likely negative impacts is also possible. Of the sixteen OECD countries, eight did so, with more to follow. Funds also flow via **multilateral aid**, via sources such as the World Bank, co-ordinated by the Committee of International Development Institutions on the Environment (CIDIE). Precise allocations to biodiversity are difficult to isolate and screening of all projects for impacts was slow to develop, with the World Bank introducing **environmental impact assessments** (EIAs) in 1989.

Clashes between different parts of the World Bank still occur. The CIDIE was also the first administrator for the Global Environment Facility. The GEF suffered initial problems. Projects were classic protectionist conservation, with little feel for wider contexts, local participation and benefit. The lack of participation may be due to confidentiality of bank procedures. The fund also concentrated on large, short-term projects, e.g. establishment of big reserves, rather than long-term, local-scale needs in developing countries.

National Environment Funds have been touted as a way forward. Local, national and international expertise can combine. Such bodies could facilitate the flow of international funds, combining local expertise and control over what might be resented as foreign interference. There would also be accountability to donor countries and the chance to regulate the boom-time bonanzas and cash-flow starvation that can afflict conservation efforts.

Overall estimates in 1994 for protected area budgets were US$4.1 billion, a lot of money but far short of the US$1 trillion for defence, US$245 million on agricultural subsidy and a long way short of the US$17 million that the Global Biodiversity Strategy estimates as the annual need for effective management of global biodiversity.

Tropical forest conservation and sustainable use

The 1980s focus on tropical forestry led to the foundation of two schemes in 1983: the **Tropical Forestry Action Plan (TFAP)**, combining the UN and conservation NGOs, and the **International Tropical Timber Agreement (ITTA)**, a cartel of producer and consumer countries, now constituted as the International Tropical Timber Organisation (ITTO). Although seen as rivals, especially given the potentially different agendas of their founders, their actions have been similar, including protection and transfer of skills and benefits. The TFAP (1985–91) co-ordinated aid for the conservation and sustainable use of forests, based on National Forest Management Strategies for each country, channelling aid into recognised projects. An estimated US$8 billion was distributed to projects in over seventy countries, with biodiversity conservation taking 8–9 per cent of the budget. However, the TFAP is widely regarded as a failure; it was narrowly focused on commercial exploitation of foresters and in many cases appeared to accelerate the industrialisation of forest destruction. The ITTA had the hallmarks of a commodity agreement but from the start recognised conservation as part of utilisation to maintain ecosystems and biosphere. The total ITTA budget in 1991 was US$20.4 million, of which the committee funding forest management plans, the mechanism used to build in conservation, had US$12.3 million. The ITTA is now on its third incarnation and continues to promote both commercial exploitation and sustainable management of forests. The various ITTAs have funded US$247 million worth of projects between 1987 and 2003. In 2003 the budget of US$14.4 million contained US$6.2 million aimed at reforestation and forest management, the remainder going on market analysis and industry development.

Following the vacuum left by failure of the TFAP the UN has now led development of the much more widely based Collaborative Partnership on Forests (CPF), co-ordinated by the UN's **Food and Agriculture Organization (FAO)**, which draws together organisations such as the World Bank, ITTO and Secretariat for the Convention on Biodiversity, which has been in place since 2001 and drawing on existing funding sources such as the GEF. The CPF is intended to develop the full range of sustainable forest uses and conservation. The significance of forest as more than just unharvested wood is explicit: the FAO recognised not only the significance of forest biodiversity for food, shelter and ecosystems services but also their less tangible ethical and cultural importance. Global deforestation continues; the FAO estimates that global net annual losses ran at 0.22 per cent. Table 5.1 gives estimates of 2003 timber stocks, production and loss rates for countries and globally.

Debt for nature swaps

Debt for nature swaps convert part of a country's financial debt into a domestic obligation to conservation. Precise details vary but the main steps are broadly similar. Debt for nature swaps seldom involve simply wiping out a debt in return for a promise to conserve. A conservation organisation must raise money to buy the debt from the bank to which it is owed. Since the debts of so many developing countries are never likely to be paid off, the conservationists can buy the debt at a fraction of its real value, often 15–30 per cent of the amount owed. The debt is bought in a hard currency. The conservationists can then negotiate with the debtor country to fix a rate at which the hard currency can be converted into local currency, called the **redemption price**. The debtor country then issues a **bond in local currency** covering the redemption price. The bond can be used to finance projects. All parties stand to gain. The banks get some repayment, the country is extricated from some debt, conservation projects can be established.

Table 5.1 Extent, use and loss of forests from around the world. The estimate of production includes all types, so timber for export and fuelwood primarily for local use. Data from FAO

	Country area, 000 ha	Forest area, 000 ha	Forest as % of total land area	Annual net change 1990–2000	Annual timber production, 000 m³	Timber volume 000 m³
Africa as a whole	2,878,394	649,866	21.8	–0.8	611,266	46,472,000
Cameroon	46,540	23,858	51.3	–0.9	12,866	3,211,000
Zimbabwe	28,685	19,040	49.2	–1.5	9,828	765,000
UK	24,160	2,794	11.6	+0.6	20,138	359,000
Brazil	845651	543,905	64.3	–0.4	272,473	71,252,000
Indonesia	181,157	104,986	58	–1.2	141,894	8,242,000
World as whole	13,063,900	3,869,455	29.6	–0.2	4,470,480	386,352,000

Too good to be true? Swaps can be see as an imposition with foreigners controlling a country's debts but since the debts are originally owed to foreign banks and the swaps result in local control of finance, this is less of a danger. Swaps can risk imposing their ideals and projects but sensitive handling can create a positive link helping local communities. More awkward are problems with governments loath to see power and available money disseminated at a local level. Debt swaps may set off unexpected market forces, creating local inflation. At the international level the prospect of debts being bought out has raised their redemption value. As more and more are purchased, the value of those left rises. Redemption values of several South American countries' debts may have rallied for no other reason than the intervention of conservationists, making debt swaps more expensive. So far debt for nature swaps have redeemed less than 10 per cent of the debt burden, risking a false impression of success whilst unsolved debt crises continue to drive the ultimate human pressures that are the major threat to biodiversity.

Protected areas

Protected areas is the catch-all term for what many think of as nature reserves. The shifting emphasis of conservation away from purely protection of wild nature has stretched the definition of reserves. The Rio Convention defines a protected area as a geographically defined area which is designated or regulated and managed to achieve specific conservation goals, the Global Biodiversity Strategy as legally established land or water area under public or private ownership that is regulated and managed to achieve specific conservation goals. These broad definitions recognise the variety of purposes conserving genetic, species or ecosystem diversity and, increasingly, the importance for human welfare and culture, for sustainable use and integration of multiple uses such as recreation and harvesting. What unites protected areas is regulation of human impacts, within a defined area, established by legislation or ownership and with specific goals. As UNEP wryly puts it, 'packing devices to achieve defined ends'.

Table 5.2 Historic examples of protected areas and allied laws. Note the similarity to modern themes via establishment of areas and regulation of activities

Protected area	Purpose
Middle Kingdom Egypt	Hunting licences to regulate wildfowling
Sumero-Babylonia laws 1900 BC	Fines for cutting down trees
Wessex Law Code (Britain) AD 700	Penalties for cutting down trees
Norman Britain (eleventh century)	'Forests', a specific designated for hunting districts exclusive to the king
Henry VI (1423–27)	Laws to regulate netting of river fish and control damage to coastal marshes and sea banks
Victoria (1860–92)	Eight Acts of Parliament dealing with game licences, protection for wild birds and poaching
Yellowstone National Park, USA (1872)	Widely regarded as first modern protected area

This approach to conservation is ancient, in sacred groves, hunting parks and wilderness. Themes redolent nowadays are apparent in these ancient schemes (Table 5.2 lists examples of historic protected areas), the desire for wilderness, perpetuation of wildlife, regulation of exploitation and protection of forests. The modern concept of reserves was a nineteenth-century evolution of these strong foundations.

The purpose and role of protected areas

The use of protected areas for conservation recognises that habitat loss, fragmentation and degradation are major causes of extinctions. The long-term survival of many species and ecosystems outside of protected areas in landscapes exploited by people is unlikely. We do not have the expertise or resources to maintain most species ex-situ, in zoos, botanical gardens or gene banks. We do not even know how many species there are, so it is hoped that protecting areas will include the unknown as well as familiar. Reserves also nurture aesthetic and cultural attraction, a wild idyll of landscape, often the spur to conservation centuries before the concept of biodiversity was coined. Protected areas are a fundamental device for conservation in all countries, though the numbers, area and goals vary with landscape and culture. Historically protected areas have arisen by a serendipity of social mores, money and available land. The recognition of the true extent of biodiversity, the hidden iceberg of species numbers, genetic variety, ecosystem services and future insurance submerged beneath the historical focus of a few game species and myths of wilderness, has resulted in attempts to distil the goals of reserves and strategies for selection, design and management.

Protected areas used to be seen as isolated pockets in the wider, exploited landscape, pockets that could somehow be managed to retain their original character with little regard to the world beyond. But wire fences cannot exclude the economic pressures, global change, pollution or local people. Effective use of protected areas now recognises the importance of ecological processes and change. Reserves should not be set up which are too small, fragmented or isolated so that their ecology is not sustainable. Reserves must account for evolutionary processes and genetic variety. Reserve management also commonly includes use by people.

Protected areas for biodiversity commonly cite four main goals for conservation:

- functioning ecosystems, recognising that processes that are being guarded may themselves drive change
- biodiversity, e.g. species richness, endemics, distinctive wildlife, often requiring management to prevent change to maintain the desired state
- species specific, typically charismatic, flagship species
- exploitation and sustainable use, recognising human benefit now or future potential.

Selection, design and management

Selection, design and management are linked. Selection results from the strategy, the ideals, the techniques used. Design is a local response to available choices. Management is the plan for the future.

Selection

The threats to biodiversity have prompted a more rigorous approach to selecting reserves than the accidents of history. Various approaches combine **area** and **representativeness**. Ideally national boundaries are less important than selection within a **bioregion**, though this may be politically impossible. A larger area will harbour more species and more communities, and have more robust ecological processes. Any selection strategy should aim to include the full range of taxa and habitats in a country. Selecting representative areas requires some **classification** of ecosystems. Selection by area and representativeness should then be reinforced by identification of **gaps**, especially important for a country with historic reserves commonly the result of an ad-hoc focus on rare and special species and habitats often ignoring examples of the common. In addition **hot spots**, defined by species richness, endemism or distinctiveness, can be added. Selection for ecologically viable area, representativeness, filling gaps and hot spots presumes that expertise and data exist and reflect the intrinsic value of biodiversity. Other criteria can be useful. Reserves selected for **flagship species**, typically large vertebrates, are common. An area sufficient in resources and large enough for a viable population of animals like this protects many other species, even if little known. The symbolic value and perhaps tourist income add political and financial benefits. **Indicator taxa**, usually species readily found and identified but known to reflect wider biodiversity, can be useful where data are limited. **Wilderness**, defined almost by our lack of knowledge, certainly by lack of human impact, has been used at a global scale. Three global wildernesses have been proposed: first, an arc through the Amazon, second the Congo basin, third, the New Guinea rainforests. Areas this size provide global services and an evolutionary pool. **Utility** may be used to safeguard sustainable exploitation or future potential, e.g. conserving wild ancestors of crop species as an insurance. Even if very little is known of a region, the **ecological principles** outlined in Chapter 3, particularly the patterns associated with broad primary factors of area and gradient, may be used to designate sites.

Design

Protected area design has been a fraught topic. In developed countries any chance for design is limited by the scramble to protect remnants whatever their size and shape. Design from scratch is possible only when carving out new areas in wilderness. Design has become snarled up in a debate between theory, largely based on **island biogeography theory** since reserves are effectively islands in a sea of degradation and practice. Some sound basic principles do exist.

Size and number

A larger area will often be a more successful reserve, harbouring more biodiversity. Larger populations are more resistant to extinction and larger species often require wider ranges. The amount of edge and fragmentation will be less relative to the core and the whole will be less vulnerable to a single natural disturbance (at least of a kind characteristic to the habitat such as fire or flood). Essentially the area must be large enough to be a **minimum dynamic area** for the ecology. The size must be sufficient to provide the resources required to support the **minimum viable population** of a species, be it rarity, flagship or keystone. So 'big is best' is a simplification. All depends on the size of species and nature of habitat in question. The size needs to be big enough to avoid accidents that are demographic (e.g. not finding a mate or being eaten) or environmental (e.g. fire, flood), genetic or social meltdown and to permit migration and ecosystem function.

An equal area divided between several small reserves may be more appropriate than one big site, the **Single Large Or Several Small (SLOSS)** debate. Several reserves may be less at risk than one larger site hit by disease, predation or severe disturbance because the risk is divided. This is especially so if the ecology of a species is out of synch between different reserves. For example if one stage of a species' life cycle is vulnerable to disease or disturbance, populations in all reserves will be not be at this same stage. Surviving reserves act as recolonisation sources. A scatter of reserves may hold more species in total because they can cover more variety of habitats. Genetic diversity can be maintained. Size and number vary with the goals of the protected area.

Shape, fragmentation, buffers and zoning

The shape of a reserve will determine the length of **edge** relative to area. Extensive edges have been seen as bad, allowing protected species to leave the safety of reserves, entry for exotics, human pressures and pollution. **Fragmentation** increases edge length and creates other problems. Populations may be cut up into smaller, less viable pockets. Barriers to dispersal can be either physical (e.g. fences) or behavioural, such as open roads that animals will not cross. Circular, intact areas have been promoted as the ideal, minimising edge to area. Sprawling shapes, especially if long, can cover more habitats increasing diversity. Edge habitats can also be valuable, either due to intrinsic combinations of species or as foraging habitat. The greater problem is that the edge of any reserve is not invulnerable to human impacts.

Design incorporating the insights of the SLOSS debate and shape should see a protected area within the wider landscape context. Reserves can be established in **sets**, recognising distances over which species can move. **Buffer zones** can be created. Many protected areas cannot be isolated from the surrounding land, bringing problems such as agricultural fertiliser run-off in Britain or local people grazing livestock and cutting wood in Africa. New reserves may have to accept historic use and future pressures. Increasingly reserves are created accepting multiple uses and with **zoning** in place of abrupt edges. Zoning allows different activities within defined parts of the reserve. This also helps build an infrastructure to suit the reserve. Unesco Man and Biosphere reserves use this approach with a core area totally protected, a surrounding buffer permitting traditional use and an outer transition zone allowing sustainable development and experimental management. This approach recognises problems rather than taking an oppressive, conflict stance.

Corridors and connections

Linked to size and shape are ideas for routes, whether open corridors, linear habitats, stepping stones or flyways to help movement between reserves. Dispersal includes daily routines, annual migration, dispersal stages and shifting overall range. Corridors have been promoted to permit recolonisation of isolated reserves and extend the effective ranges of smaller areas containing large, mobile species. Corridors also bring threats by allowing spread of disease and exotics and by providing a focus for poaching and increasing edge. As so often, the answers are species specific.

Position in landscape

Reserves cannot be designed in isolation from the surrounding land or seascape; they must be seen as part of the wider ecology, including human land uses. A regional approach can incorporate sets of reserves, buffers, zones and multiple use from the start. Part of the success of a protected area will be the effectiveness of management beyond its boundaries, requiring a national and regional coherence as well as local actions such as education (see Box 27).

Box 27

Reconnecting and restoring

The realisation that fragmented nature reserves may be increasingly ineffective has begun to affect protected areas in the developed and developing world. In Britain two ambitious schemes to extend, reconnect and restore isolated East Anglian fenland reserves have been proposed. The Great Fen Project is intended to restore 3000 hectares of farmland which will link together the existing Woodwalton and Holme Fen Reserves. Both are National Nature Reserves and SSSIs. Woodwalton is also a SAC and Ramsar site, designated as a good example of natural wetland and for the rich invertebrate and plant life. Both are stuck in a landscape of intensive arable farmland. The project began in 1999 with the aim of purchasing the intervening land, connecting the sites, allowing restoration and providing a buffer to maintain the original sites' ecological health, which provides an important floodwater buffering capacity to the surrounding landscape. The first purchases of farmland were made in 2002, adding 40 per cent to the Woodwalton Site. A similar scheme has been proposed to extend Wicken Fen, Britain's first protected area, adding at least 3700 ha to the existing 320 ha site, again allowing restoration of an effective fenland landscape. The first land purchases were made in 2000. Both schemes have emphasised local benefits, primarily recreation, social inclusion and tourism-based jobs in the region with the United Kingdom's fastest growing human population, although this is still a largely top-down consultation. Not all the public have been so keen to see new wetlands creeping up to their back gardens following recent concerns about flooding and climate change in lowland England.

Management

The best selection and design protocols will be a waste of time without effective **management goals**. Protected areas need clear goals and effective management to realise them. This is a matter of political will, expertise and money. Areas with unclear or conflicting goals will fail. Management for one species or habitat may threaten others. Unclear responsibilities and conflict can disrupt progress. Willingness to adapt in response to research and experimental management, perhaps to alter the size and shape of a reserve, can be important. Involvement of local communities in decision-making, financial benefits and management of costs such as damage to crops or dangerous wildlife can all ameliorate pressures.

Protected area problems

Not all protected areas are successful. Management failures have occurred on many sites. Very successful management can also create threats by imparting a false sense of security and providing symbols of success that none dare alter. This attracts increased pressure from tourists, poachers, local people and industry, all keen to exploit pockets of wildlife. Local conflicts are commonplace where protected areas try to operate an exclusion policy. Shifts in conservation policy, e.g. emphasising benefits to and involvement of local people, have highlighted the lack of sustainable use of biodiversity resources inside many reserves. Utilisation may be difficult to build into existing management plans. There are problems of bias of designated sites, globally and nationally. Some ecosystems are extensively protected, others underrepresented and establishment of marine reserves lags behind terrestrial habitats. Illegality can create effective protected areas. The very rare Lears macaw, thought restricted to 177 individuals at one site in South America, has been discovered at a second, high quality habitat site. Bird trappers and traders have left the birds alone for fear of drug cartels growing marijuana around the site.

The extent of global protected areas

The IUCN maintains the World Database on Protected Areas (WDPA) which provides an inventory of the number and extent of protected areas (see Figure 5.4). The first such database was drawn up in 1962 on behalf of the UN and listed over 1000 sites. By 1997 this had increased to 12,754 and in 2003 stood at 102,102 protected areas. This increase is partly due to changes to the eligibility criteria. Prior to the 2003 list, sites smaller than 10 sq. km on land or islands of less than 1 sq. km as well as all sites lacking an official IUCN classification of their purpose were excluded. The latest lists have removed both these restrictions, recognising that small sites may be valuable and that the lack of IUCN categorisation was excluding over 34,000 sites from lists. Nonetheless, the pattern of rapid increase in numbers of sites is a real trend. Since 1997 the IUCN has adopted a seven-category classification, though it is still possible to find sites described under an earlier eleven-category system (see Box 28).

Numbers of protected area reserves and allied designations globally and in Europe and Africa are given in Table 5.3 (see also Plate 24).

Besides the classic problems of underfunding, poor management and constant threats of exploitation the increasing co-ordination of global reserve resources has shown big differences of biogeographical coverage. Tables 5.4 and 5.5 show the variation of extent based on broad regional and biome categorisations. Table 5.5 suggests that the extent of protected areas has increased markedly throughout most biomes, but this is largely an artefact of smaller sites and unclassified sites, which had previously been excluded from assessments.

Ex-situ conservation

Protected areas are the main device for in-situ conservation, relying on ecosystems to sustain themselves, with some help. An alternative for endangered species is to establish

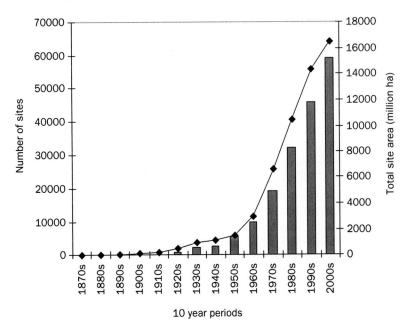

Figure 5.4 Global extent of protected areas measured by number and extent. The data are summarised into ten year periods, using all IUCN categories and begin with the Yellowstone Park in 1872 through to 2004. Total area is shown as columns, number of sites by the line

Source: Data modified from http://sea.unep.wcmc.org/wdbpa/statistics/.

a captive population away from its natural habitat. This ex-situ approach includes **captive breeding**, **release** schemes and **biodiversity banks** such as seed collections. Several species have famously been saved from extinction by ex-situ schemes. Some species still survive only in captivity. The earliest known example of an ex-situ success is Père David's deer, extinct in the Chinese wilds some 3000 years ago but surviving in a park. Other celebrity examples include the Ne Ne goose of Hawaii, nurtured at the Wildfowl and Wetlands Trust, the black footed ferret in the United States, the entire remnant population taken into captivity between 1987 and 1991 to escape canine distemper, and the *Partula* snails of Hawaii, a rare example of popularised invertebrate conservation due to the survivors living in sandwich boxes in London Zoo.

Ex-situa and in-situ conservation should not be thought of as alternatives or rivals. Ex-situ schemes should be linked to in-situ projects to augment remaining populations or establish new ones. Ex-situ schemes can link directly with protected areas, providing stock to release, research data and, in the case of zoos, funds from ticket sales. Ex-situ populations are a refuge whilst in-situ threats can be dealt with. Captive stock create additional benefits as popular icons, inspiring interest and finance. They act as flagships and are important for education, spotlighting endemism, hot spots or particular threats. Successful release schemes are effective symbols building more support for conservation. Wild populations will be important to captive stock to maintain genetic diversity and reinforce captive bred stocks. Ex-situ programmes raise problems equal to the benefits.

Whatever the precise number of species on Earth, 10 million, 20 million or 30 million, the resources and technology for ex-situ conservation of all of them do not exist. There is a bias to mammals, birds and some plants. Of the 20,000 terrestrial vertebrate species, estimates suggest that only 1000 species could be sustained ex-situ. Some taxa are probably impossible to maintain in captivity, e.g. the large whales. For those species that can be supported, the effort may be expensive and risky, with the money being better spent in the

---

Box 28

IUCN protected area management categories

Ia   **Strict nature reserve** Protected area managed mainly for science. Area of land and/or sea possessing some outstanding or representative ecosystems, geological or physiological features and/or species, available primarily for scientific research and/or environmental monitoring.

Ib   **Wilderness Area** Protected area managed mainly for wilderness protection.

II   **National Parks** To protect outstanding natural scenic and natural areas, including recreational and educational but not extractive use.

III   **Natural Monument/Landmarks** To protect nationally significant features of special interest or unique character.

IV   **Nature Conservation Reserve** To protect nationally significant species, communities or features requiring human intervention. Some exploitation permissible.

V   **Protected landscapes/seascapes** To maintain significant land and sea scapes characteristic of harmonious interaction of humans and nature.

The remaining categories are often separated from the true protected areas of I–V.

VI   **Resource Reserve** Relatively undeveloped areas under pressure for development. Human impact not poorly understood.

VII   **Anthropological Reserve** Natural areas where modern impact has not significantly interfered with or been absorbed by inhabitants.

VIII   **Multiple use areas** Large areas containing significant natural features and systems but also suitable for exploitation.

The IUCN also categorise sites designated by two major treaties.

IX   **World Heritage Sites**

X   **Biosphere Reserves**

The changing focus of conservation away from protectionist policies has seen the introduction of another category.

XI   **Managed Resource Protected Area** Protection of biodiversity whilst providing traditional, sustainable utilisation by local people within a largely unmodified area.

---

wild. Costs of maintaining African elephants (*Loxodonta africana*) and black rhinos (*Diceros bicornis*) in zoos may be fifty times that of protecting the same number in the wild. Nurturing tiny populations in captivity can create a false sense of security. Evolutionary and ecological variation found in separate wild populations of a species may be lost, so genetic and ecological biodiversity decline even if species diversity is maintained. Adding to captive stocks can deplete remnant wild numbers. Captive populations are typically small and scattered, risking genetic and social problems. Successful breeding brings its own problems. Captive born young may be naive, partly domesticated individuals that cannot be released; the redemption of a species in captivity may make exploitation of

Table 5.3 Numbers of IUCN category I–V reserves globally, in Europe and in Africa

IUCN management category

Region	Ia	Ib	II	III	IV	V
World	4,731	1,302	3,881	19,833	27,641	6,555
Europe	923	419	273	3,696	16,797	2,861
Africa	35	14	311	29	609	31

Source: UN List of Protected Areas 2003

Plate 24 A campsite in a Zimbabwean National Park. Permanent structures provide some refuge for tourists in an emergency but tents can also be pitched. The site is managed by the Department of National Parks and Wildlife as part of its conservation work.

Table 5.4 Extent of land surface included within protected areas across selected global regions. Regions defined by World Commission on Protected Areas, based on biogeographic characteristics

WCPA region

Antarctica	Pacific	Australia and New Zealand	Europe	East and South Africa	Central America
0.0%	2.1%	9.6%	13.1%	14.6%	24.8%

Source: UN List of Protected Areas 2003

Table 5.5 Extent of selected global terrestrial biomes within protected areas. The biomes are based on Udvardy's famous classification

Biome	Global extent (km²)	Total number of Protected Areas	% of extent within Protected Areas in 2003	% of extent within Protected Areas in 1997
Tropical humid forest	10,513,210	3,422	23.3	8.8
Temperate broadleaved forest	11,216,659	35,735	7.6	3.6
Warm desert	24,279,843	2008	10.3	4.8
Tundra	22,017,390	405	11.8	8.4
Tropical grassland	4,264,832	318	15.3	7.4
Lake systems	517,695	261	1.5	1.1

Source: UN List of Protected Areas 2003

its wild habitat politically easier once it is apparently secure ex-situ. Not all captive stocks have been successful. Some species have met their doom in captivity, e.g. the passenger pigeon, though to be fair there was only the one left. More recent schemes have foundered. The Sumatran rhino (*Dicerorhuns sumatrensis*) was considered to be so endangered in the wild that a captive stock was considered essential. Thirty-eight animals were caught, depleting wild populations, but fifteen died and none has bred. The ethics of captivity, particularly the role of zoos, has provoked criticism with suggestions that captive stocks are more to the benefit of the zoos than the animals. Extinction with dignity in the wild has been touted as preferable to miserable captivity. Even if all the objections were overcome, the ultimate problem remains. If no viable habitat is left in the wild, no release is possible.

The variety of past schemes and lessons from success and failure have driven attempts to codify principles and practice for ex-situ conservation. Ex-situ schemes are now seen as support for populations in their natural habitat. Ex-situ projects should be used as refuges for populations in immediate danger of extinction, as part of programmes to ensure the long-term survival and for education and research. Projects should promote the longevity, health and quality of life for captives, ensure that reproduction maintains stocks to avoid taking more from the wild and maintains natural patterns of genetic diversity, sex ratios and age structures. Captive breeding can then support releases, re-establish ecologically or culturally important species and perhaps bring economic benefits. These goals need an understanding of the biology of small populations, the experience, training and technologies to carry them out and management to co-ordinate work, not just with the captive stock but responsive to changes in wild populations. The result is international co-ordination of ex-situ schemes, controlling both the management of captive stocks and attentive to wider priorities for species conservation.

The IUCN/World Conservation Union has a Species Survival Commission to which the Captive Breeding Specialist Group (CBSG) reports on individual species. The International Species Information System (ISIS) uses tools such as the Animal Records Keeping System (ARKS) and studbooks which record the ancestry and whereabouts of individual animals. ISIS handles data on 4200 animals, from 395 zoos in 39 countries. There are over 300 breeding programmes, featuring over 200 endangered species. Co-ordination extends beyond cataloguing existing schemes. The CBSG and Taxon Advisory Groups use Conservation Assessment and Management Plans (CAMPs), which include Population and Habitat Viability Assessments (PHVAs) to identify priorities and future needs or opportunities. CAMPs are strategic plans for management of threatened taxa. Workshops collate existing data, assess threats and extinction probabilities and population estimates for both in-situ and ex-situ stocks. The species is then assigned to a category (critical, endangered, vulnerable or safe) and recommendations are made for action. These can include the need for PHVAs, management plan workshops, conservation ex-situ, gaps in research and technology or captive breeding. The outcome of a CAMPs review is a Global Captive Action Plan (GCAP), an international or regional plan for captive breeding but in support of in-situ conservation. GCAPs assess which species should remain in captivity, be added or need not be, and review available resources and technology plus research and financial benefits to wild populations.

Captive populations

Captive breeding or raising are central to ex-situ conservation in order to buy time for in-situ projects. Captivity can be seen as a last resort in the face of imminent extinction but ideally captive populations should be established before this, allowing time for research into species' needs and causes of decline, and to ensure that taking from the wild does not add to losses. The possibility of success, eventual release and cost compared to in-situ protection are important. Use of studbooks to track individual family trees helps minimise inbreeding and positively increase genetic diversity via planned matings. Where trans-portation of animals is physically difficult, artificial insemination, egg and embryo implants might be used.

Captive raising includes **cross-fostering**, generally incubation of eggs by domestic birds, though care must be taken so that chicks do not imprint on their surrogate parents on hatching, growing up believing they are the same species. **Head-starting** is protection of wild eggs and help for hatchlings (e.g. of turtles and crocodiles) to overcome heavy mortality at hatching, either from predation or sometimes disorientation and obstacles created by human development of beaches. Although superficially effective, such schemes cause problems. Young born in captivity may lack vital skills so that release is impossible or, conversely, they may retain the very skills and behaviours that doomed their in-situ forebears. Head-starting may be pointless since natural populations of the species involved typically withstood severe mortality of hatchlings, the threats causing population declines happening at other stages in the life cycle. There are risks even whilst in captivity. The Bali starling (*Leucopsar rothschildi*), an exquisite crested white mynah bird and Bali's only endemic species, is an endangered species, extinct in the wild and with captive stock being nurtured with the intent of restocking. In 1999 half the breeding starlings were stolen by thieves and, in a third raid in 2002, the thieves shot and wounded guards, so valuable are these starlings.

Releases

It is hoped that captive breeding leads to releases, often called **translocations**, which take several forms. **Augmentation** (or **restocking**) is the addition of captive bred or wild caught individuals into an existing population. **Reintroductions** (or **re-establishment**,

Plate 25 *Partula* snail reintroduction. Enclosure built on the Pacific Island of Moorea to protect reintroduced *Partula* stocks. The electrified walls are intended to exclude predatory *Euglandinia* snails. Reintroduced *Partula* survived initially and bred but *Euglandinia* got in over branches falling across the walls wiping out the Partula. Photograph Dave Clarke/London Zoo

restoration and **repatriation**) establish new populations within a species' historic range, in areas from which it has been lost (see Plate 25). This could be back into original habitat from which the captive stock was taken or elsewhere in the range if suitable sites exist. **Introductions** are releases outside the original range, which can be a threat to the host ecosystem. Wanton introductions have been a major cause of extinctions as exotic species flourish. Even those undertaken for conservation have been pilloried as unnatural gardening. International translocations of vulnerable species, such as attempts to establish black rhinos in Australia beyond the reach of poachers, are acceptable examples.

Releases can bring many benefits. The likelihood of extinction in the wild is reduced, ecosystem functions can be restored, especially if the species is a keystone, and the risk of catastrophe and inbreeding to captive stocks are ameliorated. Successful schemes can act as potent flagships for conservation. However, releases are almost certainly doomed if the original causes of decline and extinction remain. Given the long, expensive and difficult nature of releases, some principles have been developed from analysis of successful schemes.

The release area

The original cause of decline and any new threats must be eliminated. The ecosystems should not have changed so much, even if the changes are totally natural, that released animals could not survive. Releases work best into top quality habitat, in the core of the historic range. The habitat should be protected and large enough to support a viable population, allowing for an increase in numbers to reach this. Habitats with likely competitors, predators or at the carrying capacity for remaining wild populations of the release species should not be used.

Release stocks

Translocated wild stock generally fares better than captive bred, but wild stock should not be used if this endangers the original population. Captive stock should be trained to cope, either by human handlers or by interacting with surrogates or remnant local populations. The released animals should not endanger wild populations by importing disease or genetic and social disruption. The release of large numbers over many years is preferable. There is some evidence that for birds and mammals success rates increase as numbers released rise to 100, but beyond that there is little gain.

Support

Soft releases, providing support in the initial stages with food, shelter and familiarisation, work better than abrupt (hard) releases. Techniques for support and monitoring should be properly developed beforehand, combined with knowledge of species' biology to assess chances of success. Sufficient funds should be available not only to set up a release but also for long-term support. Releases should accord with local laws, which should be amended to embrace released stock in protective regulations.

Public relations

Public support is vital through education and involvement. A supportive public will be an asset. This may involve incentives or compensation if the released species is perceived as a pest, though any releases likely to endanger the public can generally be discouraged. Relevant government departments and NGOs should all be involved with clear lines of responsibility established.

Estimates of the value of release projects vary with the criteria used to judge success. A review of releases of threatened animals in Australia, New Zealand, Canada and the United States between 1973 and 1986 estimated that only 44 per cent succeeded. A stricter review of releases from 1900 to 1992, anywhere in the world, with criteria of captive bred stock, reaching self-sustaining populations of 500 plus, estimated only 11 per cent success.

Biodiversity banks

Ex-situ conservation uses biodiversity banks, starting with zoos and botanic gardens, keeping live animals and plants through to banks of genes, stored seeds and microbes.

Zoos

Zoos began as menageries for status, entertainment and education but since the mid-1960s they have consciously adopted a role in conservation, as refuges for the endangered and sources of captive supply for releases. London Zoo has been dubbed the 'Ark in the Park', playing to this theme with advertisements in the early 1990s picturing endangered animals with the caption 'Without zoos you might as well tell these animals to get stuffed'.

The actual success of zoos has been fiercely contested. The World Conservation Monitoring Centre accepts the 1990 *International Zoo Yearbook* survey listing 878 zoos globally, in 83 countries, with an inevitable bias to the developed world, 298 in Europe versus 16 in Africa. Taxa held are also biased towards charismatic vertebrates. The 878 zoos listed hold 1,232,226 specimens of vertebrates but most were of little importance for conservation. Of 629 mammals that the IUCN recognised as globally threatened, the zoos held 20,628 specimens, scarcely 10 per cent of the total of 202,000 mammals held. Only 10 per cent of the zoos were reckoned to make a significant contribution to conservation; 20 full species and a similar number of subspecies have been saved by conservation in zoos. Only 9 threatened mammals have captive populations greater than 500. However, the anti-zoo lobby creates a one-sided picture. For example the critical Zoo Inquiry cited only 405 zoos out of the 10,000 contributing to ISIS, ignoring many species conservation programmes not included under the ISIS scheme. Criticisms of zoos also ignore their greatest impact; estimates of global visitor numbers to zoos in the 1990 *International Zoo Yearbook* were 619 million, over one-tenth of the world's population. The role of zoos both in species survival and education is tremendous; I am sure that childhood visits to London Zoo were one of the mightiest influences encouraging my interest in wildlife. Table 5.6 summarises some zoo successes and problems (see also Plates 26 and 27).

The World Zoo Conservation Strategy (WZCS) in 1993 estimated there were 10,000 zoos globally of which 1000 could take part in conservation work. The WZCS sees zoos as complementary to, not a substitute for, other conservation work. There are five main aims:

- to identify the areas where zoos can make a contribution to the conservation and sustainable use of natural resources
- to develop support and understanding for conservation by zoos from authorities and other social and political organisations
- to convince zoo authorities that the greatest purpose they serve is the contribution they can make to conservation
- to assist zoos in the formulation of policies including conservation priorities
- to augment contributions of individual zoos via contacts within the global zoo and conservation network.

Table 5.6 Numbers of individuals held in zoos versus numbers actually born in zoos for five charismatic mammals

Taxon	In captivity	Number born in zoo captivity	Notes
Sumatran rhino, *Dicerorhuns sumatrensis*	13	1	Famous example of much criticised capture programme
Giant panda, *Ailuropoda melanoleuca*	9	3	The most charismatic mammal of all and subject of furious debate
African hunting dog, *Lycaon pictus*	269	0	Has bred in captivity outside of zoos
Gorilla, *Gorilla gorilla*	617	300	Successful zoo breeding
Golden lion tamarin, *Leontopitheucs r. rosalia*	592	Most	Much cited success story leading to release

Source: Groombridge 1992

Plate 26 Jersey Wildlife Preservation Trust training programme. The JWPT International Training Centre works to share and spread skills for conservation, supporting local staff from the developing world. Here a trainee from the island of St Lucia is working with the Curator of Reptiles. Photograph Tony Allchurch/JWPT

Plate 27 The Penguin Parade, Edinburgh Zoo. Zoos remain a source of popular entertainment. Perhaps the impact of these experiences upon our awareness and affection for wildlife is every bit as important as the self-conscious conservation programmes many zoos now headline

Critics of zoos, led by groups such as Zoo Check, pointed to zoos' discovery of conservation as a ticket-selling exercise, with many of the 1000 potential zoos poorly managed, the criteria for their inclusion being simply membership of a recognised zoo association rather than any real commitment of facilities. The involvement of known bad zoos, transfers of captive breeding stock as part of legitimate projects to poor zoos and zoos as sources of surplus stock to the pet trade and laboratory animal market worsened the problems for good zoos. In Britain the crisis came to a head in 1991. London Zoo, threatening imminent closure due to lack of funds, brought in a Chinese giant panda, Ming Ming, at a time when China was accused of a rent-a-panda philosophy, depleting wild stocks because captive breeding could not meet demand. Unfortunately Ming Ming flew in when London Zoo's male was in Mexico, so a German panda was brought over. To seal the nuptials Cilla Black, host of hit TV show *Blind Date*, presided. They did not breed. Soon afterwards the running of London Zoo was restructured in an attempt to focus more on conservation. Zoos are an easy target and their critics missed the point. The potential of zoos within integrated conservation programmes is vital. Attacking the good because of the sins of the worst is no help to conservation.

Zoos can have an important role in conservation and the glare of negative publicity has forced serious consideration of how this can be done. **Education** is a key task. Zoos are largely dependent on visitors for their income but this can be turned into a positive asset, so that visitors realise their entrance fee has contributed to conservation work as well as learning from their visit. Zoos can concentrate on a limited number of **flagship species**, charismatic or keystone, primarily vertebrates, with the intention to supply release programmes in protected areas. A zoo could redeploy funds and space to build a viable ex-situ population and expertise. The release would protect habitats and other taxa ex-situ. Zoos also remain a last refuge for species reduced to one or two vulnerable wild populations. This integration of in-situ and ex-situ conservation is particularly encouraging but brings its own problems. The sheer number of zoo animals of little conservation value is a burden and they cannot be summarily disposed of to free resources and space for target species. London Zoo discovered this in the wake of the changes when rumours of unwanted animals to be put down caused public concern. Zoos in the developed world are vital as **training centres**, not only for zoo staff but also for allied conservationists, especially from the developing world. Training can help to transfer skills and education, and above all act as a boost to the self-confidence of staff who often work in thankless isolation. Training should include follow-up contacts to keep up the momentum. The most famous example is the International Training Programme in Conservation and Captive Breeding of Endangered Species, based at Gerald Durrell's famed Jersey Wildlife Preservation Trust. This programme is seen as part of integrated conservation, including research, in-situ and ex-situ work. Increasingly zoos have focused on specific projects, playing to their strengths in animal husbandry and working with partners abroad, rather than just promoting everything the zoo does as innately 'conservation'. Currently London Zoo has six global conservation priorities: deserts and rangeland (e.g. the reintroduction of the Arabian oryx), bushmeat and forests (the zoo was instrumental in setting up the UK Bushmeat Campaign), marine and freshwater (e.g. Project Seahorse), carnivores and people (in particular tigers), island ecosystems and UK native species. The last two are good example of zoos realising that **invertebrate conservation** is viable. The limited space required, speed of recovery and time to release and public fascination for the bizarre but economically and ecologically important invertebrates all help. Schemes such as the *Partula* snails (see Plate 23) and British field cricket (*Gryllus campestris*) breeding programmes at London Zoo have become national news items. The animals may be small and obscure but their plight is dramatic. In March 2004 London Zoo took delivery of fifty-six Bermudan land snails in a desperate effort to establish a refuge population safe from the ravages of introduced predatory snails and ants in Bermuda. **Aquaria** have an additional role, refining extensive commercial aquaculture expertise for conservation and acting as a buffer against the impact of the aquarist fish trade.

Botanic gardens

The first modern botanic garden was founded in Padua in 1545, to collect together the ever-increasing variety of plants brought back by explorers and act as an inventory and resource bank. There are over 1500 botanic gardens worldwide, including 532 in Europe and 82 in Africa. They harbour 35,000 species, 15 per cent of the world's plant species diversity. The Royal Botanic Gardens, Kew, alone holds 27,000. Some 800 botanic gardens claim to be active in conservation, though problems include underfunding, poor maintenance, taxonomic bias (orchids, succulents and temperate trees are over-represented) and often tiny populations, perhaps just one show specimen. Even the best botanic gardens have collections skewed towards the robust and ornamental, rather than systematically reflecting the full range of endangered plants. Botanic Gardens Conservation International co-ordinates data and expertise across 327 gardens and many national networks exist. The Botanic Gardens Conservation Secretariat of IUCN co-ordinates priorities for collecting and identification of under-represented groups and establishment and training for staff in the developing world. There has been a shift from diverse collections of exotic species to specialisation on taxa, region or plants of particular use. Additional benefits from botanic gardens are historic records such as herbaria, education and links to protected areas. A few species exist solely in botanic gardens, e.g. the birch tree (*Betula murrayana*) is down to one specimen.

Like zoos, botanic gardens have increasingly focused on key conservation priorities. Kew, in London, highlights the following key roles: the UK scientific authority for CITES, home of the millennium seed bank (aiming to safeguard 24,000 species from around the world and currently claiming to have secured nearly all UK flowering plants), the Threatened Plants appeal (channelling funds to species-specific projects), training in conservation and the People and Plants Project, an international network to bring together scientific expertise and traditional knowledge to promote sustainable environments.

Seed, **microbe** and **gene banks** are hi-tech storage facilities. Seed banks are effective for conservation of genetic diversity in many plants. The number of seeds stored represents a viable population taking up very little space. Many can be stored long term, for a hundred years or more, and seeds capture the genetic spectrum of a species, compared to a handful of individuals. Storage at low humidity (5–8 per cent) and temperatures (–10 to –20°C) is effective for many seeds. However, non-seeding plants or those producing **recalcitrant** seeds which must germinate rapidly or die cannot be stored. Some 20 per cent of seed plants are recalcitrant, including many tropical and crop species. Even longer lasting **orthodox** seed gradually loses viability and requires growing on to reset seed. Initial set-up costs for storage equipment can be high and vulnerable to breakdown. There were some 528 seed banks linked to botanic gardens in 1990, often specialising by region or taxon. Particular attention has been given to improving banks of wild seed of domestic species and their relatives. Fifty major banks are linked via the Consultative Group on International Agricultural Research; the Centre for Plant Conservation published guidelines for conservation of endangered plant genetic biodiversity in 1990. The guidelines suggest:

- The order of priority is species in danger of extinction, unique taxa, those that can be reintroduced, those with ex-situ potential and those with potential value.
- Seed should be collected from five populations per species from throughout their geographical and environmental range.

- Seed should be collected from ten to fifty individuals per population.
- The numbers of seeds taken must take account of viability.
- Seeds should be collected over several years so as not to impact on local populations.

Field gene banks are an allied technique relying on cultivation of plants in managed sites. Although perhaps within the plant's natural range, more intensive ex-situ management is used. Field gene banks exist for many crop and forestry species and are especially useful to maintain local landraces in the conditions which spawned them. **Gene (or germplasm) banks** store examples, called accessions, of the genetic diversity of species. Thirteen International Agricultural Research Centres have been funded to concentrate on domestic species and their relatives and now hold over 90 per cent of landraces of main crop species, e.g. Centro Internacional de La Papa, Lima, Peru, is mandated as a centre for potatoes, with 500 varieties of potato, 1500 of wild potatoes and 5200 of sweet potato. Plants have also been conserved in clone banks, using artificial propagation of cuttings and tissues. Seed and gene banks have been the subject of tension over ownership, control and access, especially between source (developing) and recipient (developed) countries.

Microbe banks have grown from co-ordination of existing cultures. The World Federation of Culture Collections accredits 345 cultures from 55 countries which are important as pure sources for research. Their origins have resulted in a bias to disease organisms and a few useful microbes such as nitrogen-fixing bacteria. UNEP's promotion of Microbial Resource Centres (16 in 1992) has widened their role to include conservation, collection, training and secure holding facilities.

Conservation of animal genetic diversity has relied less on banks. An FAO/UNEP Animal Genetics Resources Programme was launched in 1982 primarily for domestic breeds, to develop a database and techniques such as cryopreservation of semen, eggs and embryos. A World Watch List of endangered breeds includes information on such technologies.

The digital zoo

For many people in the developed world, natural history television has been one of the most popular genres of broadcasting, a source of knowledge and inspiration that has been a major influence, creating an awareness of the variety of life on Earth and the threats so many species face. The wealth of footage and imagery has inspired the creation of an electronic menagerie, ARKive, intended to create a lasting audio-visual record of life on Earth, a 'Noah's ark for the internet era'. ARKive is a project of Wildscreen, a UK charity that grew out of the work of the BBC Natural History Unit in Bristol. The aim of the project is to compile an audio-visual library of all the species currently listed as threatened by the IUCN, 12,000 at present. This will provide both a scientific record and an haunting testimony to the state of planet Earth in the third millennium. ARKive not only uses well-known broadcast material but also attempts to find any private film, video or sound before this is forgotten or destroyed, and makes this accessible over the internet. An internet video may not be what we think of as a biodiversity but it is certainly the nearest you or I will get to seeing many of the creatures described in this book and is already providing an evocative inventory combining some of humanity's most powerful technology and most fragile neighbours.

Mars calling? Mu Arae beckons

In August 1996, as I finished writing the first edition of this book, the news was full of reports that scientists have identified possible evidence of ancient life from Mars. The evidence consisted of what might be microfossils of microbial life in a meteorite fragment found in Antarctica. (A meteorite? Recovered from the polar ice? Had these scientists never seen that film *The Thing*?) If the patterns in the rock were once alive, the surface of Mars now appears utterly dead. Huge meandering gullies suggest that water once flowed over the surface but no longer. If there was life on Mars it is either long gone or buried deep. Either way the news from Mars, which in 2004 included film from planetary rovers of what appeared to be sedimentary rocks on a now barren desert slope, makes the richness of life on Earth seem even more precious. It also opens up new horizons for biodiversity on an interplanetary scale. We would do well to remember the destructive history of human expansion over this planet before we set off for any others, not least, as I write this second edition, when astronomers have announced the discovery of the first super-Earth planetary system a mere fifty light years away around the mu Arae star.

Conservation is rife with problems and challenges but this should not eclipse what we do know. The principles and practice of good conservation are well understood throughout the world. The real challenge is whether we can be bothered. Ultimately conservation is not about the details of genetics, zoo management or treaty clauses. Conservation depends on our commitment. Or else, should mu Arae or anywhere else harbour inquisitive aliens, all they might ever see of the fantastic and beautiful life with which we share the Earth are ARKive's Ark of the Extinct.

Summary

- Conservation covers species and habitat protection, sustainable use of biodiversity, including ensuring that those doing the conserving reap the benefits plus treaties and funding.
- International and national legislation is used to control biodiversity trade, protected areas and financial aid.
- Protected areas recognise habitat loss as a major cause of degradation. They range from totally protected reserves to multi-use zones.
- Ex-situ conservation includes captive breeding, zoos, botanic gardens and gene banks. Increasingly ex-situ work supports in-situ conservation within protected areas.

Discussion questions

1   If ivory trading can guarantee the survival of the African elephant, should moral concerns still result in an ivory trade ban?
2   What is the strategy for and structure of protected areas in your country?
3   Should commercial companies be allowed to copyright and own natural genetic biodiversity?

Further reading

See also

Local economic incentives and conservation, Chapter 4 pp160–164.

Ecological causes of extinction, Chapter 4 pp138–142.

The road to Rio: the conceptual history of biodiversity, Chapter 1 pp5–19.

General further reading

McNeely, J.A. (ed.) (1994) Protected Areas (special issue). *Biodiversity and Conservation*, 3(5).

Meffe, G.K. and Carroll, C.R. (1994) *Principles of Conservation Biology*. Sinuaer Associates, Sunderland, MA.
Comprehensive textbook exploring conservation policy and practice in the light of biodiversity and changing concepts.

Miller, K. Allegretti, M.H., Johnson, N. and Jonsson, B. (1995) *Measures for Conservation of Biodiversity and Sustainable Use of its Components*. In V.H. Heywood (ed.) Section 13 *Global Biodiversity Assessment*. Cambridge University Press, Cambridge.
Detailed review of approaches to conservation including protected areas, legislation, economics and ex-situ conservation.

Olney, P.J.S. Mace, G.M. and Feistner, A.T.C. (eds) (1994) *Creative Conservation*. Chapman and Hall, London.
Principles and case studies of captive breeding and reintroduction schemes.

Primack, R.B. (1993) *Essentials of Conservation Biology*. Sinauer Associates, Sunderland, MA.
Comprehensive textbook exploring conservation policy and practice in the light of biodiversity and changing concepts.

Biodiversity and Conservation. 1995. Special issue in tribute to Gerald Durrell.
Discussion of the role of zoos.

Other resources

UK conservation programmes

Desmoulin whorl snail Biodiversity Acton Plan: http://www.ukbap.org.uk/UKPlans.aspx?ID=629
Fen Habitat Action Plan: http://www.ukbap.org.uk/UKPlans.aspx?ID=18
Great Fen Project: http://www.greatfen.org.uk/

Ex-situ conservation

ARKive: http://www.arkive.org
Kew Gardens: http://www.rbgkew.org.uk
London Zoo: http://www.zsl.org/conservation/

 # Glossary

aerobic
Environment with or metabolic process requiring oxygen.

allele
Different forms of a particular gene.

allogenic
Biological patterns or processes arising outside the ecosystem which they eventually affect.

allopatric
Spatially separate populations. Allopatric speciation occurs when a previously continuous population is split, each subpopulation diverging sufficiently to become different species.

anaerobic
Environment lacking or metabolic process occurring in absence of oxygen.

anagenesis
New species arising by changes to an existing species through time so that ancestors and descendants are sufficiently different to warrant status as different species. Also known as **chronospeciation** or **phyletic evolution**.

Archeabacteria
One of two groups of bacteria (the other are **Eubacteria**) comprising the prokaryotes. Characterised by diverse tolerances to extreme environmental conditions. Broadly equivalent to domain Archaea.

assemblage
Collection of species living together but without any special associations, mutual dependence and indifferent to precise combination of species present.

augmentation
Addition to individuals to reinforce existing population.

autogenic
Biological patterns or processes arising within the ecosystem which they eventually affect.

bequest value
Valuation of biodiversity based on not using a resource now but estimate of potential worth to future generations.

biocontrol
Control of a pest species by use of another organisms, e.g. disease, parasite or predator.

biodisparity
The degree of difference between components of biodiversity, e.g. biodisparity between a haddock and a snail is greater than that between two species of fish.

biodiversity
Totality of genetic, taxonomic, ecosystem and domestic richness and variation within the living world.

biogeography
Study of biological patterns and processes with a spatial and often historical context.

biome
Global or continental scale ecosystem zones defined by vegetation and fauna largely determined by climate.

bioregion
Coherent natural area defined by landscape and species. Contrasts to regions defined by artificial political boundaries.

biosphere
The planet's land, sea and air ecosystems characterised by the presence of life.

biosphere people
Affluent populations able to use economic power to exploit markets from around the world. Divorced from intimate ties to local environment.

breed
A recognised, standard form of a domestic animal species.

CAMPFIRE
Communal Areas Management Programme For Indigenous Resources, a Zimbabwean scheme to link conservation of wildlife with economic benefit to local people.

Centre of Plant Diversity (CPD)
Areas particularly rich in plant life which, if protected, would safeguard the majority of wild plants of the world. Defined by **IUCN**.

chemoautotrophy
Metabolic oxidation of simple inorganic compounds in environment to release energy from to power manufacture of organic food compounds.

chimera species
Domestic hybrids created from separate species which do not routinely interbreed and retain features from both parents.

chloroplast
Cell **organelle** that is site of photosynthesis.

chronospeciation
New species arising by changes to an existing species through time so that ancestors and descendants are sufficiently different to warrant status as different species. Also known as **anagenesis** or **phyletic evolution**.

cladogenesis
Speciation by splitting of ancestral lineage into several surviving lineages, each sufficiently different to warrant status as separate species.

Clovis overkill
Losses of north American **megafauna** around 12,000 years ago attributed to spread of Clovis hunter culture down through the Americas.

community
Collection of species living together with implication of strong interdependence and resulting patterns sensitive to precise combination of species present.

contingent valuation
Valuation of biodiversity based on amount of money people say they are willing to pay, or forgo, towards conservation.

cross-fostering
Use of surrogate, usually domestic, species to incubate eggs of threatened species in captivity.

cultivar
A recognised, standard form of a domestic plant species.

cyanobacteria
Prokaryote allies of true bacteria with ability to photosynthesise, often called blue green algae.

dambo
Zimbabwean name for small river valley wetlands found throughout much of southern Africa.

debt for nature swap
Purchase of part of a country's foreign debt by a conservation organisation who establish conservation projects within the debtor nation in return for waiving debt payments.

deep time
The hundreds and thousands of millions of years of geological and evolutionary history.

DNA (deoxyribonucleic acid)
The molecule containing the genetic code held within the nucleus of a cell.

domain
Classification of life based on analysis of ribosome RNA. Results suggest that all life forms belong to one of three domains, Bacteria, Archaea or Eucarya.

EBA (Endemic Bird Area)
Sites characterised by high numbers of endemic or restricted range bird species, defined by International Council for Bird Preservation, as part of global analysis of endemicity hot spots.

ecological redundancy
The idea that there are more species present in many ecosystems than are required to maintain proper ecosystem function.

ecological refugees
Peoples displaced from traditional lifestyle intimately linked to local environment but lacking **biosphere people's** economic security and often forced to move into new territory.

ecosystem
The living, biotic, inhabitants and their physical, abiotic, environment.

ecosystem function
The physical outcome of species' activity within an ecosystem typically referring to cycling of chemicals or alteration of the physical environment, e.g. photosynthetic production of oxygen.

ecosystem people
Traditional cultures closely wedded to local biodiversity and resources, their economy based on muscle power with rights and responsibilities often maintaining sustainable use of environment.

ecosystem services
The benefits to life, including humanity, accruing from some ecosystems functions.

Ediacara
Late Pre-Cambrian fauna, named after Edicara Mountains in Australia where first fossils were found.

endemic
A species found within one country, or sometimes other defined area, e.g. bioregion, and nowhere else.

ethnobotany
Traditional folk classifications of plants typically based on medicinal, food and symbolic use.

Eubacteria
The modern bacteria plus cyanobacteria, one of two groups of bacteria (the other are **Archeabacteria**) comprising the **prokaryotes**. Characterised by diverse tolerances to extreme environmental conditions. Broadly equivalent to domain bacteria.

Eucarya
One of three domains of life, a classification based on analysis of ribosome RNA suggesting that all life forms belong to one of three domains, Bacteria, Archaea or Eucarya.

eukaryote
Life characterised by possession of discrete cell organelles, particularly a nucleus, broadly equivalent to domain **Eucarya**.

ex-situ
Conservation of biodiversity away from its natural habitat.

exaptation

An adaptation that turns out to be fortuitously beneficial for quite a different purpose from its original use.

existence value
Valuation of biodiversity based on value to a person who may never use, seen or benefit from it but is willing to financially support its continued existence.

externality
Economic cost of using biodiversity not borne by the user but avoided and often passed on so that everyone bears part of the cost, e.g. pollution, which could be treated by an industry at a cost to profits but is released, degrading everyone's environment.

FAO
Food and Agriculture Organization of the United Nations.

flagship species
Charismatic species used as focal point for conservation campaign.

GEF (Global Environment Facility)
Fund set up in 1990 and administered by World Bank to finance protection of the global commons.

gene
A discrete, heritable unit of genetic data, consisting of DNA and carrying the code to regulate a particular characteristic.

genotype
The genetic constitution of an organism.

head-starting
Captive rearing and then release of juvenile stages of a species to avoid high mortality of young.

hedonic pricing
Valuation of biodiversity based on expert opinion of economic worth.

heterozygosity
Genetic variability of individuals and populations of a species.

homozygosity
Genetic uniformity of individuals and populations of a species.

inbreeding
Reproduction within a small population of related individuals, often reducing fitness.

in-situ
Conservation of biodiversity within its natural habitat.

introduction
Release of a species into an area outside its natural range.

isotope
Forms of the atom of an element varying in numbers of neutron particles and so in mass.

ITTA (International Tropical Timber Agreement)
International agreement by cartel of tropical timber producer and consumer countries with recognition of and mechanisms for conservation projects.

IUCN
The World Conservation Union.

keystone species
Species with a fundamentally important influence on the patterns, processes and functions of their habitat.

kingdom
Twentieth-century classification of life forms based on subcellular features. This approach suggested that all life can be divided between five kingdoms, Monera, Protista, Animalia, Plantae and Fungi.

K/T boundary
The divide between the Cretaceous (standard abbreviation K) and Tertiary (standard abbreviation T) geological periods, famous for the mass extinction event that witnessed the extinction of dinosaurs.

Lazarus species
A species thought extinct but rediscovered.

lichen
Symbiotic life form, a sandwich of fungus outer coat sheltering algae.

living dead species
A species with one or some individuals still alive but inevitably doomed to extinction as these die out, leaving no progeny.

macrogenesis
Speciation by a dramatic mutation, producing significantly different offspring, often termed hopeful monsters.

megafauna
Terrestrial animals, primarily mammals, of large body size.

metapopulation
A species split into separate populations, linked by dispersal of individuals.

microfossil
A microscopic fossil, typically used to refer to fossils of microbes.

miner's canary
Species or ecosystems showing marked degradation that may be a warning sign of wider environmental damage. Analogy to canaries used by miners as early warning of gas dangerous gases.

minimum dynamic area
The smallest area required to conserve the totality of patterns, processes and functions of an ecosystem.

minimum effective population
The smallest number of individuals within a larger population that actually participate in reproduction required to ensure a species' survival into foreseeable future.

minimum viable population
The smallest isolated population required to ensure a species' survival into foreseeable future.

mitochondrion
Cell **organelle** that is site for respiration.

nucleotide base
Individual molecular building blocks that link to form DNA and RNA, their sequence determining the genetic coding.

OECD
Organization for Economic Cooperation and Development.

option value
Valuation of biodiversity based on estimates of potential uses, including insurance against the unknown.

organelle
Structures inside **eukaryotic** cells specialising in particular task, e.g. mitochondria and chloroplasts.

outbreeding
Reproduction between individuals not closely related, typically drawn from a large, heterozygous population.

parapatric speciation
Speciation as adjacent populations, stretched over environmental gradients, diverge sufficiently to warrant status as separate species.

peripatric speciation
Speciation as small, peripheral subpopulations become isolated from main population and diverge sufficiently to warrant status as separate species.

phenotype
The observed characteristics of a species, the result of expression of genotype interacting with environment.

photoautotrophy
Alternative term for photosynthesis.

photosynthesis
Trapping of light energy to power manufacture of organic food compounds.

phylogeny
Evolutionary lineage of a taxon.

phylum
Level of classification below kingdom, typically based on fundamental body form and reproduction.

polymorphic
Variable in form or character.

polyploidy
Individuals containing more than one set of genes.

prokaryote
Organisms characterised by lack of distinct cell nucleus. Equivalent to the kingdom Monera or domains Archaea and Bacteria.

protected area
Area with legal designation recognising conservation as sole or one of several purposes of the area.

radiation (evolutionary)
Evolutionary diversification, most commonly referring to increase in numbers of taxa.

rebound (evolutionary)
Recovery of ecosystem following mass extinction event.

recipient country
Country receiving raw biodiversity resource which may then be refined into useful product, e.g. tropical forest plants exported to developed world recipient countries to test for potential pharmaceuticals.

recombination
Swapping over of genes between chromosomes at cell division to produce genetic variation in gametes.

Red Data Books and Red Lists
Compilations of status of and threat to species using internationally agreed categories classifications of risk. International books and lists produced by **IUCN** have been used as models for national and regional lists.

reintroduction
Establishment of new population of a species within historic range from which it has been lost.

ribosome
Site of protein manufacture inside cell, reading the genetic code brought by RNA from original database of DNA.

RNA (ribonucleic acid)
Molecule of nucleic acids, assembled as a copy DNA genetic database, and used to transport genetic data to sites inside cells where proteins are manufactured using the information.

rRNA
RNA that forms part of structure of ribosomes.

SLOSS (Single Large Or Several Small)
Acronym for contentious arguments over best designs for protected areas.

source country
Country providing raw biodiversity resource which may then be exported to recipient countries.

speciation
Evolutionary formation of new species.

stasipatric speciation
One form of sympatric speciation wherein distinctive homozygous subpopulations become sufficiently different to warrant status as separate species.

sympatric
Populations or species sharing at least part of their range. Sympatric speciation refers to speciation without any physical separation of ranges.

taxonomy
The science of naming and classification of life.

TFAP (Tropical Forestry Action Plan)
Programme to halt destruction of tropical forests and promote sustainable use run by **FAO**.

total economic value
The value of biodiversity as the sum total of benefits from use and non-use, including assessment of value from ecosystem function and costs from externalities.

transgenic species
Species containing incorporating genes from another species, the result of genetic engineering.

trophic level
Position in food chain.

UNEP
United Nations Environment Programme.

vicariance

Scattered, disjointed distribution of a taxon due to historic dispersal via movement of continents.

WCMC

UNEP World Conservation Monitoring Centre.

WWF

Formerly the World Wide Fund for Nature (and originally the World Wildlife Fund).

zoning

Multiple use of a protected area, including not only conservation but also commercial exploitation, the uses separated between different parts of the area.

zooxanthellae

Symbiotic single-celled algae found inside corals and lost during coral bleaching events.

Further reading

General

Gaston, K.J. and Spicer, J.I. (2004) *Biodiversity: An Introduction*. Blackwell, Oxford.
Excellent introduction to the science of biodiversity.

Groombridge, B. (ed.) (1992) *Global Biodiversity: Status of the Earth's Living Resources*. Chapman and Hall, London.
The first compendium of biodiversity, still providing a wealth of data.

Groombridge, B. and Jenkins, M.D. (2002) *World Atlas of Biodiversity: Earth's Living Resources in the 21st Century*. University of California Press, Berkeley, CA.
Superb, up to date, packed with detail and beautifully produced.

Heywood, V.H. (ed.) (1995) *Global Biodiversity Assessment*. Cambridge University Press for UNEP, Cambridge.
The UN's first comprehensive assessment of biodiversity, including a great deal of data alongside detailed conceptual background materials and examples.

Howlet, R. and Dhand, R. (eds) (2000) Biodiversity: nature insight. *Nature*, 405: 207–253.
Major review and synthesis of the science of biodiversity.

Reakka-Kulda, M.L., Wilson, D.E. and Wilson, E.O. (eds) (1997) *Biodiversity II*. Joseph Henry Press, Washington, DC.
The follow-up to Wilson's (1988) classic is not as much fun but reveals how the biodiversity agenda has developed.

Wilson, E.O. (ed.) *BioDiversity*. National Academy Press, Washington, DC.
The classic publication that ignited so much work.

Wilson, E.O. (1992) *The Diversity of Life*. Harvard University Press, Cambridge, MA.
Beautifully simple and clear exploration of ecology and evolution.

The journal *Biodiversity and Conservation* should also be essential reading. Not only are individual articles relevant but also there have been many special editions covering topics such as zoos, reintroduction programmes and sustainable use of wildlife.

Websites

Convention on Biological Diversity: http://www.biodiv.org/
The Convention's own website and gateway to a wealth of information.

Department of the Environment, Food and Rural Affairs: http://www.defra.gov.uk/corporate/sdstratgey/chapter3.htm/ (March 2004)
The UK government ministry uses wildlife as one of several indicators of quality of life.

Global Biodiversity Forum: http://www.gbf.ch/

Global Biodiversity Information Facility: http://www.gbif.net/

International Institute for Environment and Development: http://www.iied.org/
  Good site showing biodiversity in a wider context of global sustainable development.

IUCN (The World Conservation Union): www.iucn.org/
  A wealth of information.

Johannesburg summit: http://www.johannesburgsummit.org/ (June 2004)

United Nations: http://www.un.org/
  Well worth checking to keep up with the continuing saga of biodiversity in the context of global politics.

United Nations Environment Programme – World Conservation Monitoring Centre. http://www. unep-wcmc.org/
  A wealth of information.

Specific topics

Origins and concepts

Ghilharov, A. (1996) What does 'biodiversity' mean – scientific problem or convenient myth? *Trends in Ecology and Evolution*, 11: 304–306.
  Biodiversity has become so ubiquitous that the concept may be unhelpful.

Harper, J.L. and Hawksworth, D.L. (1995) Preface. In *Biodiversity: Measurement and Estimation*. Chapman and Hall, London.
  Reviews origins and rise of the term biodiversity.

Lovejoy, T.E. (1980) *Changes in Biological diversity*. In G.O. Barney (ed.) *The Global 2000 Report to the President, Vol. 2 (The Technical Report)*. Penguin, Hardmondsworth.
  Lovejoy's seminal paper still captures the early themes of biodiversity research.

Marren, P. (2002) Biodiversity. In P. Marren, *Nature Conservation*.
  Chapter 11 muses, sometimes acidly, on the bureaucratic and jargon-filled world of biodiversity in the United Kingdom.

Norse, E.A. and McManus, R.E. (1980) *Ecology and Living Resources Biological Diversity*. In *Environmental Quality 1980: Eleventh Annual Report of the Council on Environmental Quality*. Council on Environmental Quality, Washington, DC.
  The second seminal paper from 1980, with notable inclusions of philosophical topics.

Wilson, E.O. (ed. (1988) *BioDiversity*. National Academy Press, Washington, DC.
  The breakthrough book.

Wilson, E.O. (1993) Is humanity suicidal? *Biosystems*, 31: 235–242.
  Wilson muses on human destruction of the environment.

Environment and nature: history and popularisation

Evans, D. (1992) *A History of Nature Conservation in Britain*. Routledge, London.

Pearce, D.F. (1991) *Green Warriors*. Bodley Head, London.

Pepper, D. (1996) *Modern Environmentalism*. Routledge, London.

Natural history filmmaking

The very influential genre of natural history filmmaking has only recently been held up for closer inspection.

Bouse, D. (2000) *Wildlife Films*. University of Pennsylvania Press, Philadelphia, PA.

Jeffries, M.J. (2000) Niche broadcasting. *ECOS* (Journal of the British Association of Nature Conservationists, 20: 70–76.

Jeffries, M.J. (2003) BBC natural history versus science paradigms. *Science as Culture*, 12: 527–545.

Mitman, G. (1999) *Reel Nature*. Harvard University Press, Cambridge, MA.

The history of life

Benton, M.J. and Twitchett, R.J. (2003) How to kill (almost) all life: the end-Permian extinction event. *Trends in Ecology and Evolution*, 18: 358–365.

Clapham, M.E., Narbonne, G.M. and Gelling, J.G. (2003) Paleoecology of the oldest known animal communities: ediacaran assemblages at Mistaken Point, Newfoundland. *Paleobiology*, 29: 527–544.

Conway Morris, S. (1998) *The Crucible of Creation: The Burgess Shale and the Rise of Animals.* Oxford University Press, Oxford.

Conway Morris, S. (2002) We were meant to be. *New Scientist*, 16 November: 26–29.

Gould, S.J. (1991) *Wonderful Life*. Penguin, London.

Conway Morris and Gould's books are as much a philosophical battle about the nature of nature as they are about ancient creatures.

Hoffman, P.F. and Shrag, D.P. (2000) Snowball Earth. *Scientific American*, 282: 68–75.

Raup, D.M. (1993) *Bad Genes or Bad Luck?* Oxford University Press, Oxford.
Extinction patterns, causes and lessons from history.

Revising our view of life's history

Alvarez, L.W., Alvarez, W., Asaro, F. and Michel, H.V. (1980) Extraterrestrial cause for the Cretaceous-Tertiary extinction. *Science*, 208: 1095–1108.
The classic article suggesting meteor impact as a cause of dinosaur extinction.

Bakker, R. (1986) *The Dinosaur Heresies*. Longman, London.
Right or wrong Bakker's interpretation of the dynastic struggles to dominate the land challenged the dusty view of life's epic history. So why was one of the baddies in *Jurassic Park 2* costumed and made up as a spitting image for Bakker?

Conway Morris, S. (1977) A new Metazoan from the Cambrian Burgess Shale, British Columbia. *Palaeontology*, 20: 623–640.
Hallucigenia makes its debut.

Margulis, L. (1993) *Symbiosis in Cell Evolution: Microbial Communities in the Archean and Proterozooic*. Freeman, New York.

Ramskold, L. and Xianguang, H. (1991) New early Cambrian animal and Onychophoran affinities of enigmatic metazoans. *Nature*, 351: 225–227.
Hallucigenia turns out (or over) to be a worm.

The current crisis

Amphibian declines (special issue). *Diversity and Distributions*, (2003), 9(2).
Synthesis of latest information on amphibian crisis.

Bellwood, D.R., Hughes, T.P., Folke, C. and Nystrom, M. (2004) Confronting the coral reef crises. *Nature*, 429: 827–883.

Eldredge, N. (1992) *The Miner's Canary*. Virgin Books, London.
Highly readable account of the current extinctions popularising the image of species loss as a warning.

Glynn, P.W. (1991) Coral bleaching in the 1980s and possible connections with global warming. *Trends in Ecology and Evolution*, 6: 175–178.

Jaenicke, J. (1991) Mass extinctions of European fungi. *Trends in Ecology and Evolution*, 6: 174–175.

Martin, P.S. (1990) 40000 years of extinctions on the 'planet of doom'. *Palaeogeography, Palaeoclimatology, Palaeoecology*, 82: 187–201.

Pechmann, J.H. and Wilbur, H.M. (1994) Declining amphibian populations in perspective: natural fluctuations and human impacts. *Herpetologica*, 50: 65–84.
Early review of amphibians as miners' canaries.

Reaser, J., Powerance, R. and Thomas, P.O. (2000) Coral bleaching and global climate change: scientific findings and policy recommendations. *Conservation Biology*, 14: 1500–1511.
Coral crisis update.

Ward, P. (1995) *The End of Evolution*. Weidenfield and Nicolson, London.
The current crisis with a historical perspective through deep time.

Species extinction and habitat loss

Alvarez, L.W., Alvarez, W., Asaro, F. and Michel, H.V. (1980) Extraterrestrial cause for the Cretaceous-Tertiary extinction. *Science*, 208: 1095–1108.
The classic article suggesting meteor impact as a cause of dinosaur extinction.

Farrow, S. (1995) Extinctions and market forces: two case studies. *Ecological Economics*, 13: 115–123.
Economics wipes out the Passenger pigeon and the buffalo herds.

Gaston, K.J. *Rarity*. (1994) Chapman and Hall, London.
Detailed and insightful coverage of rarity.

Hannah, L., Carr, J. and Lankerani, A. (1995) Human disturbance and natural habitat: a biome level analysis of a global data set. *Biodiversity and Conservation*, 4: 128–155.
Quantification of global terrestrial ecosystem losses.

Lawton, J.H. and May, R.M. (eds) (1995) *Extinction Rates*. Oxford University Press, Oxford.
Revealing, detailed and thorough collection of articles.

Ecology

Huston, M.A. (1994) *Biological Diversity: The Coexistence of Species on Changing Landscapes*. Cambridge University Press, Cambridge.

Hutchinson, G.E. (1959) Homage to Santa Rosalia, or why are there so many kinds of animals? *American Naturalist*, 93: 145–149.
Santa Rosalia's relics may be the bones of a goat but this is still a classic.

Rosenzweig, M.L. (1995) *Species Diversity in Space and Time*. M.L. Cambridge University Press, Cambridge.
  Both Huston and Rosenzweig's books are superb syntheses of the ecology of diversity but from very different perspectives, each of which hardly gets a mention in the other book.

Accessible introductions to recent ecological advances

Brown, J. and West, G. (2004) One rate to rule them all. *New Scientist*, 12 May: 38–41.

Polis, G.A. (1988) Stability is woven by complex webs. *Nature*, 395: 744–745.

Evolution

Dawkins, R. (1986) *The Blind Watchmaker*. Longman, Harlow.
  Awesome exploration of evolution's creative power.

Eldredge, N. (1995) *Reinventing Darwin: The Great Evolutionary Debate*. Weidenfeld and Nicolson, London.
  Evolutionary processes at a larger scale than the mechanics of genes.

Meyers, A.A. and Giller, P.S. (1988) *Analytical Biogeography*. Chapman and Hall, London.
  Excellent textbook, including evolutionary ecology, e.g. speciation.

Evolution has been a burgeoning topic amongst popular science writers. The following are superb books revealing both the science and the humanity underpinning evolution and ecological science.

Fortey, R. (2001) *Trilobite: Eyewitness to evolution*. HarperCollins, London.
  The history of life through the extraordinary eyes of a now extinct creature.

Hooper, J. (2002) *Of Moths and Men: Intrigue, Tragedy and the Peppered Moth*. Fourth Estate, London.
  The classic example of evolution is maybe not what it seems.

Jahme, C. (2001) *Beauty and the Beasts: Woman, Ape and Evolution*. Virago, London.
  The impact of women researchers working with the great apes, including iconic figures for conservation such as Jane Goodall and Dian Fossey.

Jones, S. (2000) *Almost Like a Whale*. Transworld, London.
  Jones rewrites Darwin's *Origin of Species*.

Weiner, J. (2004) *The Beak of the Finch*. Jonathan Cape, London.
  Beautiful tale of the research on evolution and speciation among Darwin's finches.

Ecology, ecosystem function and Planetary health

Aselmann, I. and Crutzen, P.J. (1989) Global distribution of natural freshwater wetlands and rice paddies, their net primary productivity, seasonality and possible methane emissions. *Journal of Atmospheric Chemistry*, 8: 307–358.
  Wetlands as source of atmospheric methane. Widely cited estimate of global extent of wetlands.

Chapin, F.S., III, Zavaleta, E.S., Eviner, V.T. *et al.* (2000) Consequences of changing biodiversity. *Nature*, 405: 234–242.

Glynn, P.W. (1991) Coral bleaching in the 1980s and possible connections with global warming. *Trends in Ecology and Evolution*, 6: 175–178.

Jones, C.G. and Lawton, J.H. (eds) (1995) *Linking Species and Ecosystems*. Chapman and Hall, London.
  Groundbreaking attempt to pull together different traditions of community and ecosystem research in ecology.

King, L.A. and Hood, V.L. (1999) Ecosystem health and sustainable communities: north and south. *Ecosystem Health*, 5: 49–57.

Loreau, S.M., Naeen, S. and Inchausti, P. (eds) (2001) *Biodiversity and Ecosystem Functioning: Synthesis and Perspectives*. Oxford University Press, Oxford.
Excellent synthesis of recent advances.

McCann, K.S. (2000) The diversity–stability debate. *Nature*, 405: 228–233.

Pimental, D., Westra, L. and Noss, R.F. (2000) *Ecological Integrity: Integrating Environments Conservation and Health*. Island Press, Washington, DC.

Williamson, P. (ed.) (1995) Integrating Earth System Science (special issue). *Ambio*, 23(1).
Review articles of global geochemical cycles, their importance and the politics of wise management.

Biodiversity inventory

Biodiversity hotspots: http://www.biodiversityhotspots.org/

Francesco, P. (2002) *Biodiversity and Natural Product Diversity*. Pergamon. Amsterdam.

Gaston, K.J. (2000) Global patterns of biodiversity. *Nature*, 405: 220–227.

Gaston, K.J. and Hudson, E. (1994) Regional patterns of diversity and estimates of global insect species richness. *Biodiversity and Conservation*, 3: 493–500.
Good example of research into global species total.

Gewin, V. (2002) All living things, online. *Nature*, 418: 362–363.

Hawksworth, D.L. (ed.) (1995) *Biodiversity, Measurement and Estimation*. Chapman and Hall, London.
Progress, possibilities and problems of inventory work.

Olsen, D.M. and Dinerstein, E. (1998) The Global 200: a representative approach to conserving the Earth's most biologically valuable ecoregions. *Conservation Biology*, 12: 502–515.

Purvis, A. and Hector, A. (2000) Getting the measure of diversity. *Nature*, 405: 212–219.

Stork, N.E. (1993) How many species are there? *Biodiversity and Conservation*, 2(2): 215–232.
Good example of research into global species total.

Wilson, E.O. (2003) The Encyclopaedia of life. *Trends in Ecology and Evolution*, 18: 77–80.
The new drive to describe all taxa.

Wrigley, S.K., Hayes, M.A., Thomas, R., Chrystal, E.J.T. and Nicholson, N. (2001) *Different approaches to describing the variety of life on Earth; molecular diversity. Biodiversity. New leads for the pharmaceutical and agrochemical industries*. Royal Society of Chemistry, London.

Microbial diversity

Finlay, B.J. (2002) Global dispersal of free-living microbial eukaryote species. *Science*, 296: 1061–1063.

Kerr, R.A. (2002) Deep life in the slow, slow lane. *Science*, 296: 1056–1058.

Torsvik, V., Overas, L. and Thingstad, T.F. (2002) Prokaryotic diversity – magnitude, dynamics and controlling factors. *Science*, 296: 1064–106.

Wetlands

Aselmann, I. and Crutzen, P.J. (1989) Global distribution of natural freshwater wetlands and rice paddies, their net primary productivity, seasonality and possible methane emissions. *Journal of Atmospheric Chemistry*, 8: 307–358.
Wetlands as source of atmospheric methane. Widely cited estimate of global extent of wetlands.

Dambo Research Unit (1987) The use of dambos in rural development, with reference to Zimbabwe. OA final project report R3869, file ref. ENG 505/512/10 ODA, London.

Ramsar Convention on Wetlands: http://www.ramsar.org/
The global wetlands treaty.

Rodwell, J.S. (ed.) (1995) *British Plant Communities. Vol. 4. Aquatic Communities, Swamps and Tall Herb Fens*. Cambridge University Press, Cambridge.
British National Vegetation Classification for many wetland types.

SADC Wetlands information system: http://www.sadc.int/wetlands
Background on Southern African wetlands.

Taylor, A.R.D., Howard, G.W. and Begg, G.W. (1995) Developing wetland inventories in southern Africa: a review. *Vegetatio*, 188: 57–79.
Overview of wetlands in southern Africa, including Zimbabwe.

Conservation concepts

Adams, J.S. and McShane, T.Q. (1996) *The Myth of Wild Africa: Conservation without Illusion*. University of California Press, Berkeley, CA.

Adams, W.M. and Mulligan, M. (eds) (2003) *Decolonising Nature: Strategies for Conservation in a Post-colonial Era*. Earthscan, London.
Attitudes to nature, both in the developing and developed world.

Department of National Parks and Wildlife Management (NPWLM) (1992) *Policy for Wildlife*. NPWLM, Harare.
'Wildlife is a unique economic resource'.

Gibson, C.G. (1999) *Politicians and Poachers: The Political Economy of Wildfire Policies in Africa*. Cambridge University Press, Cambridge.

Hulme, D. and Murphree, M. (2001) *African Wildlife and Livelihoods: The Promise and Performance of Community Conservation*. James Currey, Oxford.
A wealth of examples, case studies and background.

Marren, P. (2002) *Nature Conservation: A Review of the Conservation of Wildlife in Britain 1950–2001*. HarperCollins, London.
The recent UK experience. Key issues from an insider out of love with the conservation bureaucracy.

Woodcock, K.A. (2002) *Changing Roles in Natural Forest Management: Stakeholders' Roles in the Eastern Arc Mountains, Tanzania*. Ashgate, Aldershot.
Negotiated conservation strategies as an improvement on participatory schemes.

Conservation practice

Several excellent textbooks bringing together developments in conservation, in part reflecting the rise of biodiversity, have been published since the mid-1990s.

Hunter, M.L. (1996) *Fundamentals of Conservation Biology*. Blackwell, Oxford.

Laird, S.A. (ed.) (2002) *Biodiversity and Traditional Knowledge: Equitable Partnerships in Practice*. Earthscan, London.
Superb compendium of this growing field.

Meffe, G.K. and Carroll, C.R. (1994) *Principles of Conservation Biology*. Sinauer Associates, Sunderland, MA.

Primack, R.B. (1993) *Essentials of Conservation Biology*. Sinauer Associates, Sunderland, MA.

Sutherland, W.J. (ed.) (1998) *Conservation Science and Action*. Blackwell Science, Oxford.

Websites

Biodiversity and Protected Areas searchable database: http://earthtrends.wri.org/searchable_db/index.cfm?theme=7/ (July 2004).

UN Protected Areas Programme database:
http://www.unep-wcmc.org/protected_areas/index.html/ (July 2004).

Economics for conservation

Anderson, T.L. and Hill, P.J. (eds) (1995) *Wildlife in the Market Place*. Rowan and Littlefield, Lanham, MD.
Examples and background to sustainable use of wildlife as a conservation policy.

Balmford, A., Bruner, A., Cooper, P., Costanza, R., Farber, S., Green, R.E., Jenkins, M., Jefferies, P., Jessamy, V., Madden, J., Munro, K., Myers, N., Naeem, S., Paavola, J., Rayment, M., Rosendo, S., Roughgarden, J., Trumper, K. and Tuner, K. (2002) Economic reasons for conserving wildlife. *Science*, 297: 950–953.

Edwards, V.M. (1995) *Dealing in Diversity: America's Market for Nature Conservation*. Cambridge University Press, Cambridge.
You can shoot African big game on ranches in the United States. Here's why.

Global Environment Facility website: http://www.gefweb.org/

Richardson, J.A. (1998) Wildlife utilisation and biodiversity conservation in Namibia: conflicting or complementary objectives? *Biodiversity and Conservation*, 7: 549–560.

Swanson, T. (1999) Conserving global diversity by encouraging alternative development paths: can development coexist with diversity?, *Biodiversity and Conservation*, 8: 29–44.

Global forest conservation

Food and Agriculture Organisation of the United Nations: http://www.fao.org/biodiversity/Forests_eco_en.asp (July 2004)

International Tropical Timber Organisation (ITTO): http://www.itto.or.jp/

People and Plants International: http://peopleandplants.org/

United Nations Forum on Forests: http://www.un.org/esa/forests

Extinctions and threats

Bushmeat campaign: http://bushmeat.net/

de Merode, E., Momewood, K. and Cowlishaw, G. (2002) The value of bushmeat and other wild foods to rural households living in extreme poverty in Democratic Republic of Congo. *Biological Conservation*, 118: 573–581.

Rao, M. and McGowan, P.J.K. (2002) Wild meat use, food security, livelihoods and conservation. *Conservation Biology*, 16: 580–583.

Sperling, L. (2001) The effect of civil war on Rwanda's bean seed systems and unusual bean diversity. *Biological Conservation*, 10: 989–1009.

Thomas, J.A., Telfer, M.G., Roy, D.B., Preston, C.D., Greenwood, J.J.D., Aser, J., Fox, R., Clarke, R.T. and Lawton, J.H. (2004) Comparative losses of British butterflies, birds, and plants and the global extinction crisis. *Science*, 303: 1879–1881.

Economics of biodiversity

Barbier, E.B., Burgess, J.S. and Folke, C. (1994) *Paradise Lost? The Ecological Economics of Biodiversity*. Earthscan, London.
Detail and generality well combined in a thoughtful textbook.

Constanza, R., Farber, S.C. and Maxwell, J. (1989) Valuation and management of wetland ecosystems. *Ecological Economics*, 1: 335–361.
The role and value of wetlands.

Constanza, R., D'Arge, R., de Groot, R., Farber, S., Grasso, M., Hannon, B., Limburg, K., Naeem, S., O'Neill, R.V., Paruelo, J., Raskin, R.G., Sutton, P. and van den Belt, M. (1997) The value of the world's ecosystem services and natural capital. *Nature*, 387: 253–260.
The US$33 trillion estimate.

McNeely, J.A. (1988) *Economics and Biological Diversity*. IUCN, Gland, Switzerland.

Oldfield, S. (ed.) (2003) *The Trade in Wildlife: Regulation for Conservation*. Earthscan, London.

Pearce, D. and Moran, D. (1994) *The Economic Value of Biodiversity*. Earthscan, London.
Good introductory textbook to this topic, with wetland examples.

Perrings, C., Maler, K-G., Folke, C., Holling, C.S. and Jansson, B-O. (1995) *Biodiversity Loss. Economic and ecological Issues*. Cambridge University Press, Cambridge.
Good introductory textbook.

Rayment, M. and Dickie, I. (2001) *Conservation Works . . . for Local Economies in the UK*. Royal Society for the Protection of Birds, Sandy, Bedfordshire.
The value of conservation of birds in the UK to local employment and business.

The Value of Ecosystem Services (special issue), *Ecological Economics*, (1998), 25.

Zimbabwe in detail

Child, G. (1995) Managing wildlife successfully in Zimbabwe. *Oryx*, 29: 1171–1177.

Child, G. (1996a) The role of community-based wild resource management in Zimbabwe. *Biodiversity and Conservation*, 5: 355–368.

Child, B. (1996b) The practice and principles of community-based wildlife management in Zimbabwe: The CAMPFIRE programme. *Biodiversity and Conservation*, 5: 369–398.

Duffy, R. (2000) *Killing for Conservation: Wildlife policy in Zimbabwe*. International African Institute, Harare.

Metcalfe, S. (1993) *The Zimbabwean Communal Areas Management Programme for Indigenous Resources (CAMPFIRE)*. Zimbabwe Trust, Harare.
An in-house review of the CAMPFIRE scheme from those close to its heart.
CAMPFIRE did not always work, even before recent political unrest.

Alexander, J. and McGregor, J. (2000) Wildlife and politics: CAMPFIRE in Zimbabwe. *Development and Change*, 31: 605–627.

Campbell, B.M., Sithole, B. and Frost, P. (2000) CAMPFIRE experiences in Zimbabwe. *Science*, 287: 42–43.

CAMPFIRE website: http:www.campfire-zimbabwe.org/
Now a withered relic of a once very useful site.

Dzingirai, V. (2003) 'CAMPFIRE is not for Ndebele migrants': the impact of excluding outsiders from CAMPFIRE in the Zambezi Valley, Zimbabwe. *Journal of Southern African Studies*, 29: 445–459.

Ex-situ conservation

Balford, A., Leader-Williams, N. and Green, M.J.B. (1995) Parks or arks? Where to conserve threatened mammals? *Biodiversity and Conservation*, 4: 595–607.
The economics and effectiveness of conservation in captivity versus the wild.

Hancocks, D. (2002) *A Different Nature: The Paradoxical World of Zoos and their Uncertain Future*. University of California Press, Berkeley, CA.
How zoos represent our changing views of nature, and the consequences for attitudes to the environment.

Oldfield, S. (ed.) (2003) *The Trade in Wildlife. Regulation for Conservation*. Earthscan, London.

Pearce-Kelly, P., Mace, G.M. and Clarke, D. (1995) The release of captive bred snails (*Partula taeniata*) into a semi-natural environment. *Biodiversity and Conservation*, 4: 645–663.
Partula turgida may be no more but other *Partula* snails remain the focus of intensive and thoughtful work.

Rushton, B., Hackney, P. and Tyrie, C.R. (eds) (2001) *Biological Collections and Biodiversity*. Westbury, London.

Wheater, R. (1995) World Zoo Conservation Strategy: a blueprint for Zoo Development. *Biodiversity and Conservation*, 4: 544–522.
Zoos' future role.

Websites

ARKive: http://www.arkive.org/
Kew Gardens, UK's leading Botanic Garden. http://www.rbgkew.org.uk/
CITES: http://www.cites.org/
Zoological Society of London (includes London Zoo): http://www.zsl.org/

Bibliography

Adams, W.M. and Mulligan, M. (eds) (2003) *Decolonising Nature: Strategies for Conservation in a Post-colonial Era*. Earthscan, London.

Barbier, E.B., Burgess, J.S. and Folke, C. (1994) *Paradise Lost? The Ecological Economics of Biodiversity*. Earthscan, London.

Bells, S., McCoy, E.D. and Mushinsky, H.R. (eds) (1991) *Habitat Structure*. Chapman and Hall, London.

Bibby, C.J. (1992) *Putting Biodiversity on the Map*. International Council of Bird Preservation, Cambridge.

Child, B. (ed.) (2004) *Parks in Transition: Biodiversity, Rural Development and the Bottom Line*. Earthscan, London.

Collinson, N.H., Biggs, J., Corfield, A., Hodson, M.J., Walker, D., Whitfield, S.M. and Williams, P.J. (1995) Temporary and permanent ponds – an assessment of drying out on the conservation value of aquatic macroinvetebrate communities. *Biological Conservation*, 74: 125–133.

Cooper, J.E. (ed.) (1995) Wildlife species for sustainable food production. *Biodiversity and Conservation*, 4(3): 215–219.

Cotinga (1996) Lears Macaw: a second population confirmed. *Cotinga*, February: 10.

Craig, P., Trail, P., Morrell, T.E. (1994) The decline of the fruit bats in American Samoa due to hurricanes and overhunting. *Biological Conservation*, 69: 261–266.

Dawkins, R. (1986) *The Blind Watchmaker*. Longman, Harlow.

Department of the Environment (1994) *Biodiversity: The UK Action Plan*. HMSO, London.

Edwards, V.M. (1995) *Dealing in Diversity: America's Market for Nature Conservation*. Cambridge University Press, Cambridge.

Eldredge, N. (1992) *The Miner's Canary*. Virgin Books, London.

Eldredge, N. (1995) *Reinventing Darwin: The Great Evolutionary Debate*. Weidenfeld and Nicolson, London.

Erwin, T. (1982) Tropical forests: their richness in Coleoptera and other anthropod species. *Coleopterists Bulletin*, 36: 74–82.

Farrow, S. (1995) Extinctions and market forces: two case studies. *Ecological Economics*, 13: 115–123.

Flint, M. (1991) *Biological Diversity and Developing Countries*. Overseas Development Administration, London.

Gaston, K.J. (1994) *Rarity*. Chapman and Hall, London.

Gaston, K.J. and Blackburn, T.M. (2000) *Patterns and Processes in Macroecology*. Blackwell Science, Oxford.

Gaston, K.J. and Spicer, J.I. (2004) *Biodiversity: An Introduction*. Blackwell Science, Oxford.

Gehrels, T. (1996) Collisions with comets and asteroids. *Scientific American*, 274: 34–39.

Gilpin, M.E. and Hanski, I. (eds) (1991) *Metapopulation Dynamics*. Harcourt Brace Jovanovich, San Diego, CA.

Goldsmith, F.B. and Warren, A. (eds) (1993) *Conservation in Progress*. John Wiley and Sons, Chichester.

Goodwin, H. and Swingland, I.R. (eds) (1996) Ecotourism, biodiversity and local development. *Biodiversity and Conservation*, 5(3): 275–276.

Gould, S.J. (1989) *Wonderful Life*. Hutchinson, London.

Groombridge, B. (ed.) (1992) *Global Biodiversity: Status of the Earth's Living Resources*. Chapman and Hall, London.

Groombridge, B. and Jenkins, M.D. (2000) *World Atlas of Biodiversity. Earth's Living Resources in the 21st Century*. University of California Press, Berkeley, CA.

Hambler, C. (2000) *Conservation*. Cambridge University Press, Cambridge.

Hannah, L., Carr, J.L. and Lankerani, A. (1995) Human disturbance and natural habitat: a biome level analysis of a global set. *Biodiversity and Conservation*, 4: 128–155.

Harper, J.L. and Hawksworth, D.L. (ed.) (1995) *Biodiversity: Measurement and Estimation*. Chapman and Hall, London.

Heywood, V.H. (ed.) (1995) *Global Biodiversity Assessment*. Cambridge University Press, Cambridge.

Hubbell, S.P. (2001) *The Unified Natural Theory of Biodiversity and Biogeography*. Princeton University Press, Princeton, NJ.

Hulme, D. and Murphree, M (eds) (2001) *African Wildlife and Livelihoods: The Promise and Performance of Community Conservation*. James Currey, Oxford.

Hunter, M.L. (1996) *Fundamentals of Conservation Biology*. Blackwell, Oxford.

Huston, M.A. (1994) *Biological Diversity: The Coexistence of Species on Changing Landscapes*. Cambridge University Press, Cambridge.

Hutchinson, G.E. (1959) Homage to Santa Rosalia, or why are there so many kinds of animals? *American Naturalist*, 93: 145–149.

Juste, J., Fa, J.E., Delval, J.P. and Castroviejo, J. (1995) Market dynamics of bushmeat species in Equatorial Guinea. *Journal of Applied Ecology*, 32: 454–467.

Kate, K. ten and Laird, S.A. (2000) *The Commercial Use of Biodiversity: Access to Genetic Resources and Benefit Sharing*. Earthscan, London.

Kinzig, A.P., Pacola, S.W. and Tilman, D. (2001) *The Functional Consequences of Biodiversity: Empirical Progress and Theoretical Extensions*. Princeton University Press, Princeton, NJ.

Koziell, I. (2001) *Diversity not Adversity: Sustainable Livelihoods with Biodiversity*. International Institute for Economic Development, London.

Kreuter, U.P. and Workman, J.P. (1994) Cost of overstocking cattle and wildlife ranches in Zimbabwe. *Ecological Economics*, 11: 237–248.

Lawton, J.H., Nee, S., Letcher, A.J. and Harvey, P.H. (1994) Animal distributions: patterns and process. In *Large Scale Ecology and Conservation*. Blackwell, Oxford.

Lovejoy, T.E. (1980) *Changes in Biological Diversity*. In G.O. Barney (ed.) *The global 2000 Report to the President, Vol. 2 (The Technical Report)*. Penguin, Harmondsworth.

MacArthur, R.H. (1972) *Geographical Ecology: Patterns in the Distribution of Species*. Harper and Row, New York.

McCarthy, M.A., Franklin, D.C. and Burgman, M.A. (1994) The importance of demographic uncertainty – an example from the Helmeted Honeyeater *Lichenostomus melanops cassidix*. *Biological Conservation*, 67: 135–142.

Madsen, T., Stille, B. and Shine, R. (1996) Inbreeding depression in an isolated population of adders, *Vipera berus*. *Biological Conservation*, 75: 113–118.

Margulis, L. (1993) *Symbiosis in Cell Evolution: Microbial Communities in the Archaean and Proterozoic*. Freeman, New York.

Marren, P. (2002) *Nature Conservation: A Review of the Conservation of Wildlife in Britain 1950–2001*. HarperCollins, London.

Meffe, G.K. and Carroll, C.R. (1994) *Principles of Conservation Biology*. Sinauer Associates, Sunderland, MA.

Murdoch, W.W. (1966) 'Community structure, population control and competition' – A critique. *American Naturalist*, 100: 219–226.

National Science Board (1989) *Loss of Biological Diversity: A Global Crisis Requiring International Solutions*. National Science Board, Washington, DC.

Nicolaon, K.C., Guy, R.K. and Poiter, P. (1996) Taxoids: new weapons against cancer. *Scientific American* June: 84–88.

Norse, E.A. (ed.) (1993) *Global Marine Biological Diversity*. Island Press, Washington, DC.

Norse, E.A. and McManus, R.E. (1980) *Ecology and Living Resources: Biological Diversity*. In *Environmental Quality 1980: Eleventh Annual Report of the Council on Environmental Quality*. Council on Environmental Quality, Washington, DC.

Norse, E.A., Rosenbaum, K.L., Wilcore, D.S., Wilcox, B.A., Romme, W.H., Johnston, D.W. and Stout, M.L. (1986) *Conserving Biological Diversity in our National Forests*. The Wilderness Society, Washington, DC.

O'Riordan, T. and Stoll-Kleeman, S. (eds) (2002) *Biodiversity, Sustainability and Human Communities*. Cambridge University Press, Cambridge.

Pearce, D. and Moran, D. (1994) *The Economic Value of Biodiversity*. Earthscan, London.

Perrings, C., Maler, K-G., Folke, C., Holling, C.S. and Jansson, B-O. (1995) *Biodiversity Loss: Economic and Ecological Issues*. Cambridge University Press, Cambridge.

Pimental, D., Westra, L. and Noss, R.F. (2000) *Ecological Integrity: Integrating Environment, Conservation and Health*. Island Press, Washington, DC.

Postgate, J. (1994) *The Outer Reaches of Life*. Cambridge University Press, Cambridge.

Prins, H.T.H., Grootenhuis, J.G. and Dolan, T.T. (eds) (2000) *Wildlife Conservation by Sustainable Use*. Kluwer Academic, Boston, MA.

Raup, D.M. (1993) *Extinction. Bad Genes or Bad Luck?* Oxford University Press.

Reaka-Kudla, M.L., Wilson, D.E. and Wilson, E.O. (eds) (1997) *Biodiversity II*. Joseph Henry Press, Washington, DC.

Redford, K.H. and Stearman, A.M. (1990) Forest dwelling native Amazonians and the conservation of biodiversity: interest in common or in collision? *Conservation Biology*, 7: 248–255.

Reid, W.V., Sittenfeld, A., Laird, S.A., Janzen, D.H., Meyer, C.A., Gollin, M.A., Gamaz, R. and Juma, C. (1993) *Biodiversity Prospecting: Using Genetic Resources for Sustainable Development*. World Resources Institute, Washington, DC.

Robinson, J.G. (1993) The limits to caring: sustainable living and the loss of biodiversity. *Conservation Biology*, 7: 20–28.

Rosenzweig, M.L. (1995) *Species Diversity in Space and Time*. Cambridge University Press, Cambridge.

Schulze, E-D. and Mooney, H.A. (eds) (1993) *Biodiversity and Ecosystem Function*. Springer-Verlag, Berlin.

Silvius, K.M., Bodmer, R.E. and Fragoso, J.M.V. (eds) (2004) *People in Nature. Wildlife Conservation in South and Central America*. Columbia University Press, New York.

Stone, C.P. and Stone, D.B. (eds) (1989) *Conservation Biology in Hawai'i*. University of Hawai'i, Honolulu, HI.

Udvardy, M.D.F. (1975) *A Classification of the Biogeographical Provinces of the World*. IUCN, Gland, Switzerland.

Ward, P. (1995) *The End of Evolution*. Weidenfeld and Nicolson, London.

Weddell, B.J. (2002) *Conserving Natural Resources in the Context of a Changing World*. Cambridge University Press, Cambridge.

Wheatley, N. (1994) *Where to Watch Birds in South America*. Christopher Helm, London.

Wilson, E.O. (ed.) (1988) *BioDiversity*. National Academy Press, Washington, DC.

Wilson, E.O. (1992) *The Diversity of Life*. Harvard University Press, Cambridge, MA.

Woodcock, K.A. (2002) *Changing Roles in Natural Forest Management. Stakeholders' Roles in the Eastern Arc Mountains, Tanzania*. Ashgate, Aldershot.

World Resources Institute (WRI) (1992) *Global Biodiversity Strategy*. WRI, Washington, DC.

 # Index

Note: page numbers in *italic* denote references to illustrations/plates/tables.